NEW ZEALAND'S ALPINE PLANTS INSIDE AND OUT

how New Zealand's alpine plants survive
in their harsh mountainous environment

by Bill and Nancy Malcolm

published by
CRAIG POTTON
PO Box 555, Nelson, New Zealand.
KEL AIKEN PRINTING COMPANY LIMITED
Wellington, New Zealand.

to Jean, Charlie, and Pat

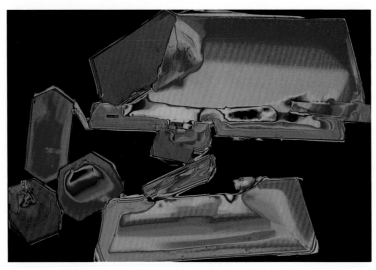

Crystals of oxalic acid

ACKNOWLEDGMENTS: The field work was greatly helped by Jean Barnhill, Tony Druce, Allan Fife, Geoff Rogers, and Pete Williams. Craig Potton and Andy Dennis commented on the manuscript, and Pete Williams read it critically. Peter Brunt of Nelson Polytechnic kindly provided sectioning and microscopy equipment. Stacey Malcolm provided technical information on sectioning, and Kevin and Susan Archer of the Forest Research Institute supplied tissue digestion enzymes.

PHOTOGRAPHS: All the plant photographs were taken by the authors with Pentax SX1000 and K1000 cameras using Fujichrome RD100 35mm film. All the plant tissue sections were made by the authors, and the micrographs were taken with an Olympus D Plan microscope fitted with polarizing optics.

PUBLISHED BY: Craig Potton, PO Box 555, Nelson, New Zealand, and Kel Aiken Printing Company Limited, Wellington, New Zealand.

BOOK PRODUCTION: Mostyn Hainsworth.

COLOUR SEPARATIONS: Robert MacLeod, Lithograph Laboratory, Wellington, New Zealand.

FILM ASSEMBLY & PLATES: Peter Dorn, Kel Aiken Printing Company Limited, Wellington, New Zealand.

PRINTING: Craig Unsworth, Kel Aiken Printing Company Limited, Wellington, New Zealand.

BINDING: Government Printing Office, Wellington, New Zealand.

DISTRIBUTION: Craig Potton, PO Box 555, Nelson, New Zealand. Telephone 054-88583.

ISBN 0-908802-04-8

contents

introduction.. 1

capturing enough nutrients............................... 11

preventing drying out...................................... 31

trapping enough sunlight.................................. 53

fending off the bugs and beasts....................... 69

gifting solutions to the next generation............ 93

glossary

index

Springtime growth in a southern heath, *Leucopogon suaveolens*.

A native orchid, *Caladenia lyallii*.

introduction

A forget-me-not, *Myosotis monroi* (of the **borage** family, the **Boraginaceae**).

Alpine plants are beautiful creatures. Their beauty is in their *structures*, like the colour and shape of their flowers, and the delicate hairs on their leaves and stems. But the structures that look beautiful to us are strictly functional to the plants, in fact utterly vital to their survival, because those structures are the solutions to the many survival problems that the plants face in their harsh mountainous home. For example, the hairs on the leaves and stem of this forget-me-not fend off leaf-eaters.

An alpine cinquefoil, *Potentilla anserinoides* (of the **rose** family, the **Rosaceae**), and Lake Sylvester in Northwest Nelson State Forest Park.

The photographs in this book show you the structures of alpine plants inside and out, and the captions below the photographs explain how those structures solve alpine survival problems. The book is organized around the problems – absorbing enough scarce vital nutrients, foiling drying winds, trapping enough sunlight, battling off the bugs and beasts, and passing on to the next generation the successful solutions to survival problems.

Most of the photographs in the book are from Northwest Nelson State Forest Park. Mining companies are interested in the minerals of the area, threatening the alpine plants that live there with a survival problem they could not cope with. Conservation groups are keen to give the area the protection of national park status.

An alpine cress, *Cheesemania latisiliqua* (of the **mustard** family, the **Cruciferae**).

Everything in the universe is quietly falling to bits, losing its structure in a relentless cosmic drift toward randomness. Nothing can escape that steady grinding loss of structure, not even living things whose very lives depend on their structure. Living things *can* escape during their life-times by re-building their damaged structure, but only if they can somehow capture enough energy and materials for the repair jobs. The trouble is, they often face problems in doing that, and an alpine plant faces more problems than most. For example, this alpine cress has little defence against hungry leaf-eating predators like hares and goats that would gladly gobble up its hard-won energy supplies. If the salads that we humans enjoy are anything to go by, browsers actually seek out the cress for its tangy taste. The cress has been badly thrashed in subalpine areas of Mt. Arthur, and it now survives only in rocky crags at high altitude, out of the reach of most predators.

One of New Zealand's showier native orchids, *Adenochilus gracilis* (of the **orchid** family, the **Orchidaceae**).

Leaf-eating predators are not the only problem an alpine plant must cope with. Even other alpine plants over-top it or else plug into its roots to steal its vital juices. As well, its harsh alpine home causes problems – the bitter cold starves it of nitrogen and phosphorus by nobbling the rot-bacteria and fungi that release nutrients from dead vegetation, and the constant wind sucks it dry of its precious water. The delicate and lovely alpine flower that delights a tramper because it looks so utterly peaceful in an alpine meadow is in fact constantly fending off savage assaults by the elements, by insects, by browsers, and by other delicate and lovely alpine flowers. It is far from defenceless, though. For example, this native orchid laces its leaves with needle-sharp crystals and poisonous chemicals – the needles punch thousands of tiny holes in the soft mouth-lining of a grazer, and then the poisons rush through the holes, causing swelling and pain.

A root-parasitic eyebright, *Euphrasia cheesemanii* (of the **figwort** family, the **Scrophulariaceae**).

Tennyson's oft-quoted line "Nature red in tooth and claw" hints of violent snarling and screeching beasts prowling some dripping tropical rainforest, but the seemingly peaceful alpine flowers speckling our frigid and barren mountains are equally violent. Rather than fight with tooth and claw, though, they wage their lethal battles with quieter strategies. Some silently fill their leaf tissues with irritants and poisons. Others squeeze out their competitors by curling their leaves and stems into dense cushions. Still others, like this eyebright, top up their nitrogen supplies by plugging into the roots of their neighbours and ruthlessly sucking out whatever is coursing through the inner pipes of their hapless victims.

An alpine cushion-plant, *Donatia novae-zelandiae* (of the **donatia** family, the **Donatiaceae**).

There is no moral "high ground" in the mountains, no right or wrong, no justice – alpine creatures survive by whatever means they are able to, and imaginative short-cuts like theft are rewarded quickly and handsomely. That startling reality threatens to spoil our enjoyment of the mountains – how can we marvel at a beautiful alpine meadow if we know that in fact all of its creatures are savagely eating and poisoning each other in a vicious tangle of lethal interactions? No problem. When he first heard a bellbird, Captain Cook was so delighted that he lyrically called its song ". . . the most melodious wild music I have ever heard, the most tunable silver sound imaginable". Two centuries later we science-wise moderns are just as charmed by the bellbird's delightful carillon of song, even though we know something Cook did not know, that the bellbird sings neither for us nor for its own joy, but to wrest and defend a territory from all comers. Cushion-plants like this one do exactly the same thing, and our knowing that they do does not diminish their beauty in the slightest.

Trampers climbing New Zealand's alpine peaks often look down from the heights they have reached to the water-logged flats where they began, and we begin our story from the same vantage ....

9

capturing enough nutrients

Rocky outcrops and scree slopes.

When we see a steep, rocky, and barren alpine landscape like this, we think that nothing can grow in it because it has no soil that plants can get their calcium, potassium, and other mineral nutrients from. With our gardening experience, we think that plants to grow well must have rich black soil full of humus. But in fact, humus is vital only as a source of nitrogen and phosphorus, and other nutrients are abundant in such rocky landscapes. They constantly leach out of the rocks because frost, ice, and rain conspire to break up the rocks into smaller and smaller chunks, which fall off and form scree slopes – and all the while the rain, frost, and even lichens and microbes **weather** the surfaces of the chunks, freeing mineral nutrients from the rock crystals.

The bristly mountain carrot, *Anisotome pilifera* (of the **carrot** family, the **Umbelliferae**), growing in the broken marble of Mt. Arthur.

Alpine plants that seed in among weathering rocks rarely suffer from any mineral nutrient shortages. However, they do suffer other problems – often they can not find enough nitrogen (which comes from rotting vegetation rather than from weathering rocks), nor can they find enough water, and their structure is constantly damaged, crushed by shifting rocks. Those problems can seriously limit their growth, even kill them.

Lake Peel on the track between Mt. Arthur and the Cobb Valley, and sphagnum moss from along its shores.

In contrast to the rocky alpine heights, the sodden water-logged flats around tarns are heavy with dead vegetation, and yet surprisingly, plants growing there suffer from a shortage of nitrogen and phosphorus. Why?– because those elements are tightly locked up by low temperature, low oxygen, and high acidity. The bacteria necessary to unlock them (by rotting the dead vegetation) simply can not survive in such conditions, and so the corpses of tarn plants like sphagnum moss pile up as **peat,** and the elements in them are released too slowly for vigorous new plant growth.

The insect-catching leaves and the flower of a common sundew, *Drosera spathulata* (of the **sundew** family, the **Droseraceae**).

Because vital elements are locked up around water-logged alpine tarns, few plants can live there, but a **carnivorous plant** like a sundew solves the problem of scarce nutrients by importing nitrogen and phosphorus in the form of insects that it lures to their doom with its many brightly refracting sticky hairs. The insects literally fly in a private supply of nitrogen and phosphorus to the sundew from nutrient-rich areas outside.

Microscope views of a sundew's insect-catching hairs with their blobs of sticky glue.

The glandular hairs on the upper surface of the sundew's leaves secrete droplets of a viscid sticky glue. The glue clings tenaciously to any foraging insects that happen to touch it, and their buzzing struggles only mire them further in a tangle of sticky hairs. Once subdued, the insects are digested within hours by enzymes secreted by the hairs. The hairs absorb the break-down products into the plant, and then drop off the empty insect carcasses.

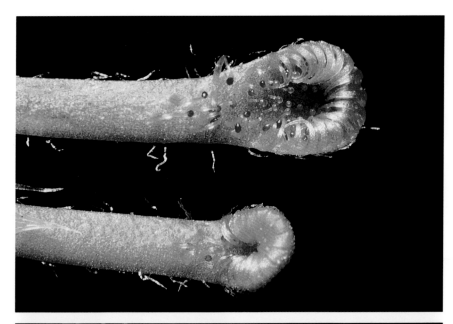

Developing leaves of the sundew *Drosera spathulata*.

The sundew's sticky glandular hairs show up early in the development of its leaves – here you see the hairs on four leaves of different ages from very young to nearly mature. When fully mature, the hairs turn bright red and their tips begin to secrete droplets of that tenacious glue as they lie quietly in ambush for their first insect victim.

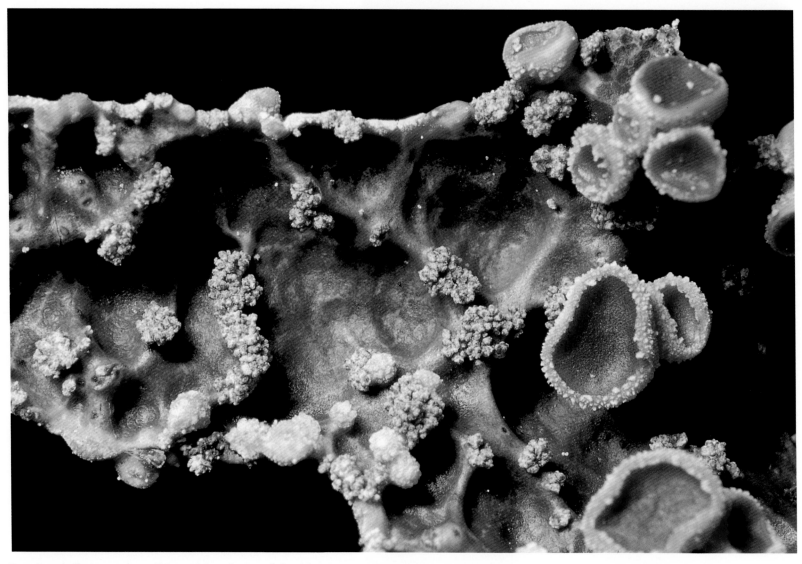

Pseudocyphellaria crocata, a lichen with a nitrogen-fixing blue-green algal partner.

Being dependent on wary buzzing insects for their supply of nitrogen is risky for sundews, a dart-board strategy that requires what we humans call luck. But many alpine lichens beat those risky odds by making their own nitrogen supply. Lichens are made up of not one but two kinds of creatures, a fungus and an alga. The lichen is actually the product of a highly complex interaction between those two creatures. This alpine lichen looks dead, but in fact is in robust good health. Its yukky colour comes from its algal partner, a **blue-green alga** called *Nostoc* that lives in a layer just under the lichen's upper surface.

19

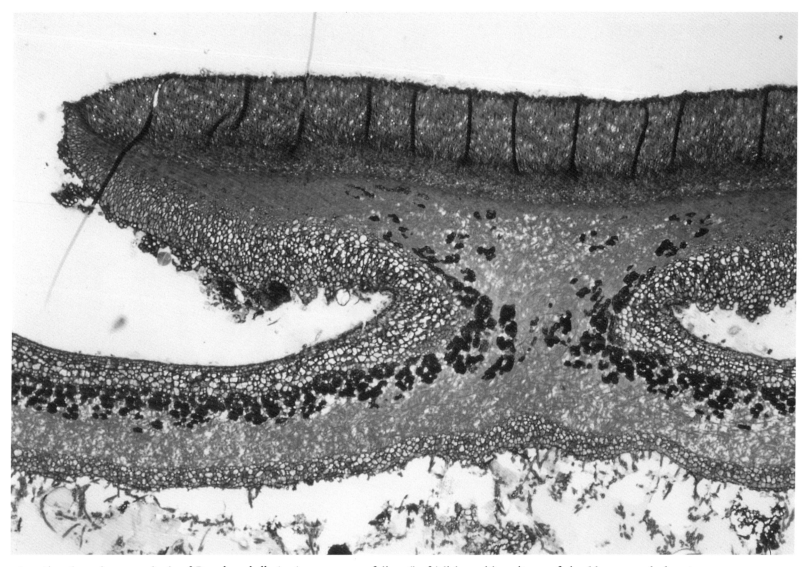

A section through a spore-body of *Pseudocyphellaria cinnamomea*, a foliose (leafy) lichen with a nitrogen-fixing blue-green algal partner.

In this section through a ***Nostoc***-containing lichen and one of its many spore-producing disks, the ***Nostoc*** alga is the dark blue layer just below the surface. In spite of being "primitive", ***Nostoc*** is a talented critter – it is able to convert chemically inert nitrogen gas from the atmosphere into soluble ammonium, hence is said to be a **nitrogen-fixer**. A lichen with ***Nostoc*** inside it reaps an enormous pay-off, because it can live in places so poor in soluble nitrogen that no other plants can survive there. The ***Nostoc*** alga handily solves the lichen's problem of capturing enough nitrogen to make the amino acids and proteins vital for building and repairing its tissues.

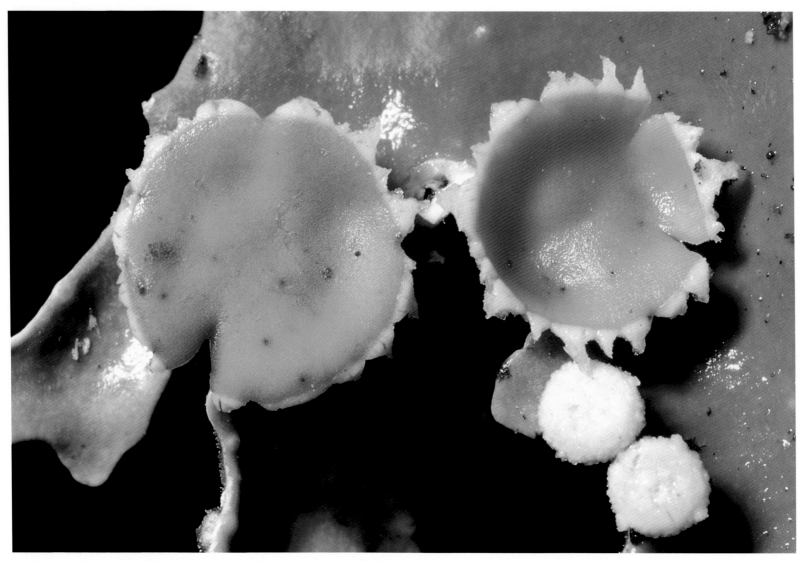

Pseudocyphellaria homoeophylla, a common foliose lichen with a green algal partner.

Because having a captive nitrogen-fixer in alpine areas is of great survival-value, it is not surprising that even lichens containing green algae (which can not fix nitrogen) often have at least some blue-green algae tucked away inside them as well. This is a close-up of one of our most common lichens, ***Pseudocyphellaria homoeophylla*** – it literally covers the ground in the beech forests on Mt. Arthur. Its lettuce-green colour tells us that it has a green algal partner.

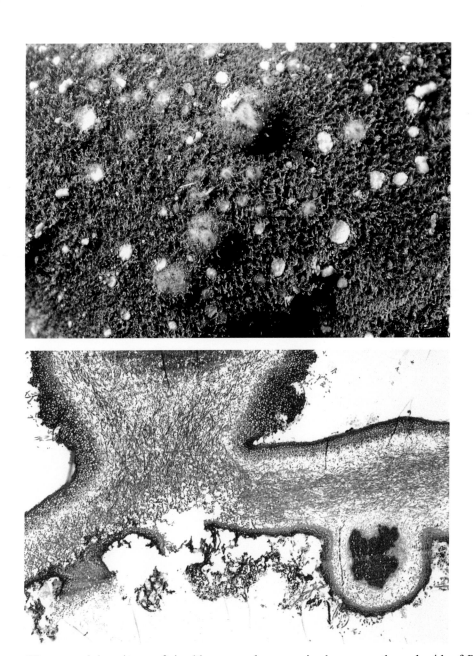

Warts containing nitrogen-fixing blue-green algae, seen in close-up on the underside of *Pseudocyphellaria homoeophylla* and in section.

But scattered over the lower surface of the lichen are wart-like bumps that contain islands of **Nostoc**. The **Nostoc** is trapped by the woolly hairs (called **tomentum**) on the lichen's underside and then carried inside the lichen, where it grows into structures that are called **cephalodia** because in section they resemble an animal's brain.

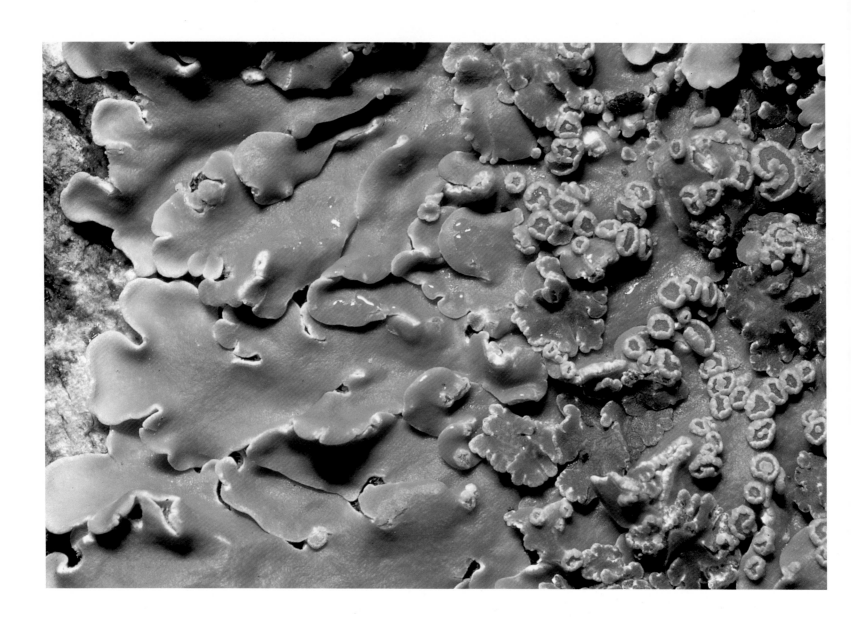

A bark-dwelling lichen, *Psoroma pholidotoides,* which has both green and blue-green (nitrogen-fixing) algal partners.

Some lichens with both green and blue-green algal partners flaunt in full view their cephalodia and nitrogen-fixing **Nostoc**, as does this **Psoroma pholidotoides**. Out in the open, the cephalodia are more efficient at fixing nitrogen, because they get more energy for the job (in the form of sunlight), and more raw material as well (inert nitrogen gas, which they passively absorb from the air around them). If the cephalodia were immersed inside the lichen instead, they would get less of both.

A snow grass tussock under attack by a root-parasitic eyebright, *Euphrasia monroi* (of the **figwort** family, the **Scrophulariaceae**).

Alpine plants face problems in getting enough nitrogen and other elements to repair the damage caused by the slow disintegration of their structure. We humans are inclined to think that those problems must always be caused by the alpines' harsh habitat, but in fact their fellow alpine creatures are often to blame. For example, alpine eyebrights like this ***Euphrasia monroi*** look harmless enough, but actually are ruthless parasites – rather than capture their own nitrogen supply, they plug into the roots of other plants unlucky enough to be growing near them, and then mercilessly suck out vital fluids. Their **root-parasitism** solves their own nitrogen problem, but it worsens their hosts' nitrogen problem.

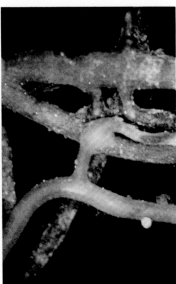

An eyebright, *Euphrasia laingii,* growing among its victims, and a close-up of a parasitic connection.

The parasitic connections between the eyebrights and their hosts are called **haustoria**. Haustoria are frustratingly small and fragile, but you can see them by digging up an eyebright in a small chunk of the vegetation surrounding it and patiently washing its tangled roots free of litter and soil in a bucket of water. You must use a hand-lens to see the haustoria, but at least there is no shortage of them – they number in the hundreds.

27

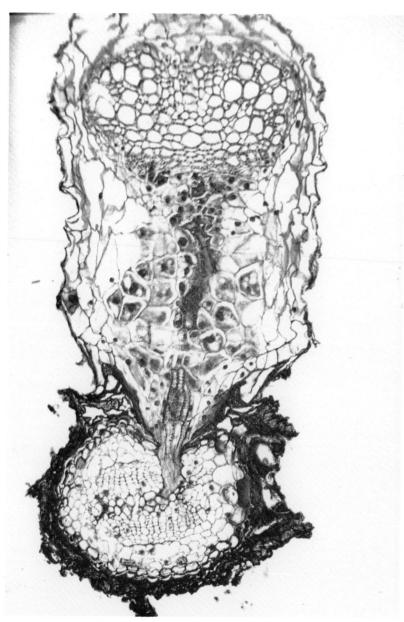

A section through the haustorium of a root-parasite.

Those many haustoria directly connect the inner piping of the eyebright and its host, giving the eyebright access to anything being transported inside its victim. Here you see a section through a haustorium, with the parasite root at the top, the host root at the bottom, and the haustorium in the middle.

A leafy flowering shoot of an eyebright, *Euphrasia laingii*.

The eyebrights are dogged parasites. They ensure that the traffic-flow through their haustoria is strictly one-way by constantly evaporating water from their leaves, thus sucking ever more fluids from their hosts. They are well-equipped to do that, too – they have far more **stomates** (pores for gas exchange) on their leaves than their hosts do.

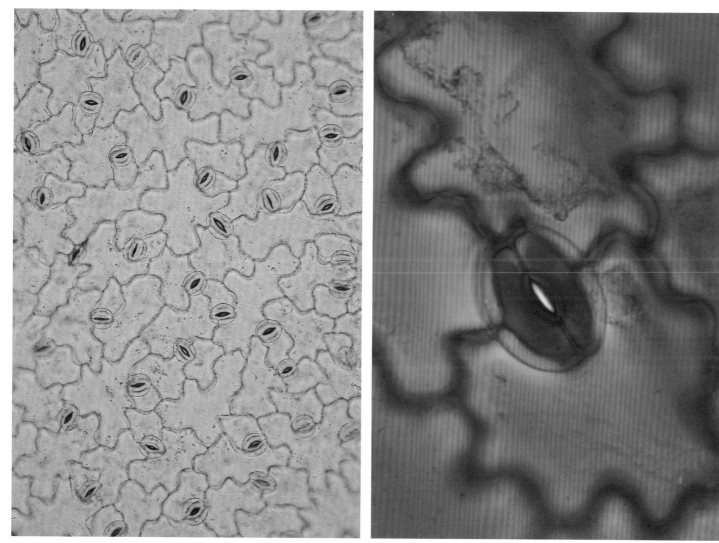

Stomates on an eyebright's leaf.

These are microscope views of an eyebright's leaf surface – what look like paving stones are the leaf's **epidermal cells**, and the doughnut-shaped structures scattered among them are the stomates. With its myriad stomates, an eyebright is a living wick that steadily evaporates away the host's precious water supply, sucking out at the same time the nitrogen-containing molecules that the parasite uses to build and repair its own structures. Given the many advantages of being parasitic, the eyebrights likely are evolving in the direction of using even more of the many free goodies that they can steal from their hosts.

Eyebrights and other alpine parasites like the mountain sandalwood and the mistletoes all can live on the cheap because they get materials and energy from their hosts rather than going to the trouble of capturing their own supplies. Like human thieves, they get rich quick, but unlike human thieves, they run no risk of being caught or even suffering the cluck-clucking of outraged moralists. They get away with anything that works, and the only measure of their success is how many offspring they can leave behind them.

preventing drying out

A shower squall on Hoary Head in the Arthur Range, and a rain-wetted flower of the Maori onion, *Bulbinella hookeri* (of the **lily** family, the **Liliaceae**).

Alpine plants must capture enough nitrogen, phosphorus, calcium, and other vital nutrients to hold their own against the ravages of a relentlessly disintegrating universe, but water is also high on their hit-list, and although it is not hard to get, it can be hard to keep. You might think it unlikely that alpine plants could ever be at risk from drying out, because mountains are wet places – no keen tramper has escaped getting soaked sometime or other.

The South Island edelweiss, *Leucogenes grandiceps* (of the **daisy** family, the **Compositae**), and the mountain spurge, *Oreoporanthera alpina* (of the **spurge** family, the **Euphorbiaceae**).

However, mountains are windy as well as wet, and their thin soils do not hold much water, so alpine plants are constantly at risk of drying out in spite of all the rain falling on them. Worse yet, they are always unavoidably losing water vapour through their open stomates. This edelweiss and mountain spurge are growing in the dry marble of Mt. Arthur.

The brown mountain daisy, *Celmisia traversii,* in its tussocky habitat, and a close-up of the dense brown hairs that give it its common name.

An alpine plant could save water during a drought by closing down its stomates, but that would shut down its vital **photosynthesis** too – after all, unless its stomates are open, it can not capture carbon dioxide from the air and make food using the energy of sunlight. So an alpine plant faces a dilemma – with its stomates open it wilts, and with them closed it starves. The alpine plant solves that dilemma by keeping its stomates open but covering them with thick hairs that slow down the wind next to its leaf surfaces. The fierce gale that we think is battering the plant in fact is only a mild zephyr right next to the stomates where a strong wind would quickly dry out and kill the plant. The stomates of this brown mountain daisy are under the dense layer of chocolate-coloured hairs that give the plant its common name.

35

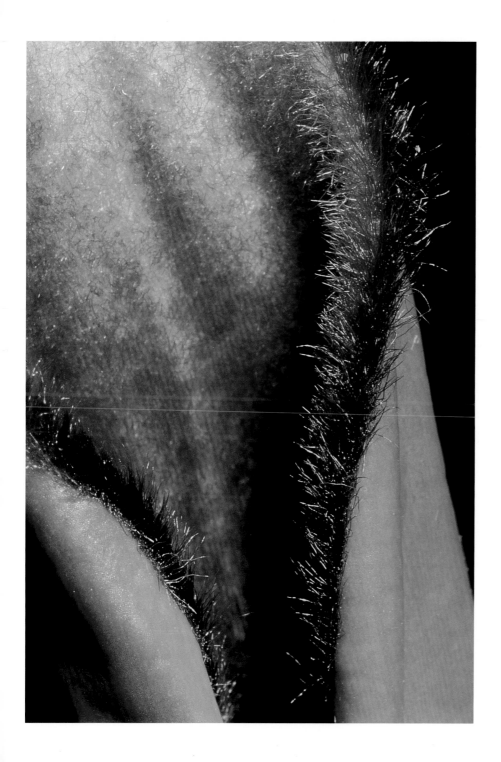

A tree daisy, *Olearia lacunosa* ssp. *lacunosa* (of the **daisy** family, the **Compositae**) in full bloom on the Mt. Arthur track, and a close-up of both surfaces of its leaves.

Dense hair (tomentum) covers the underside of the leaves of most of New Zealand's alpine daisies. In this tree daisy, the tomentum doubles the effective thickness of the leaf. Water vapour escaping from the stomates must travel through the tomentum, a barrier almost as tortuous as the leaf's inner tissues. As a result, the leaf does not lose much water even in a gale. The ***Olearia*** tree daisies take their very name from their tomentum – ***Olea*** is the genus of the olive, which has silvery tomentum on the underside of its leaves.

This species of tree daisy can be recognized easily by the shallow depressions on its leaves. Its thick outer covering (epidermis) is coated with a glossy wax that keeps out water and fungal pathogens. The stout mid-rib acts as a beam that resists bending stresses set up by the wind. The leaf depressions and the rolled margin are stiffeners, too, analogous to the patterns stamped into the flimsy sheet-metal panels of vehicles and kitchen whiteware.

A southern heath, *Leucopogon suaveolens* (of the **epacrid** family, the **Epacridaceae**).

The white stripes on the underside of the leaves of this southern heath do the same job as the fuzzy brown hairs of ***Celmisia traversii*** and the dense tomentum of the ***Olearia*** tree daisies. The white is not a paint-like pigment but thousands of tiny hairs, and the stomates are snuggled down among the hairs well out of the wind. That structural detail allows the plant to grow successfully in places that are exposed to strong winds. Said another way, the plant survives in its harsh alpine environment because it has a structure that allows it to survive there.

An alpine divaricating shrub, *Pittosporum anomalum* (of the **pittosporum** family, the **Pittosporaceae**), neatly filling crevices in the limestone of Hoary Head in the Arthur Range.

So-called divaricating shrubs like this pittosporum are common in New Zealand, and several get into alpine areas. They belong to unrelated families (the violet, madder, elaeocarp, and pittosporum families), but they look much alike with their small leaves and densely interlaced branches. Their peculiar growth form evolved as a solution to the combined stresses of wind, drought, frost, and grazing (by the now-extinct moas). Their outer branches die back from desiccation and winter ice-blast, forming a porous armour-plating that frustrates browsers and calms the wind, thus protecting the leaves growing inside the tangled mass of inner branches.

41

A large alpine willowherb, *Epilobium vernicosum* (of the **evening-primrose** family, the **Onagraceae**).

The stomates of most land plants are on the underside of their leaves. That arrangement is of greater survival-value than having stomates on the upperside, for two reasons – (1) without any stomates on it, the upperside of the leaf can be turned over completely to the vital job of intercepting sunlight, so the plant reaps a maximum energy-harvest, and (2) the underside of the leaf is cooler, so the leaf loses less water vapour, important to plants living in dry areas. But oddly enough, some of our alpine plants like this willowherb have their stomates on the upperside of their leaves, even though they live in very dry places like the cracks in limestone rocks.

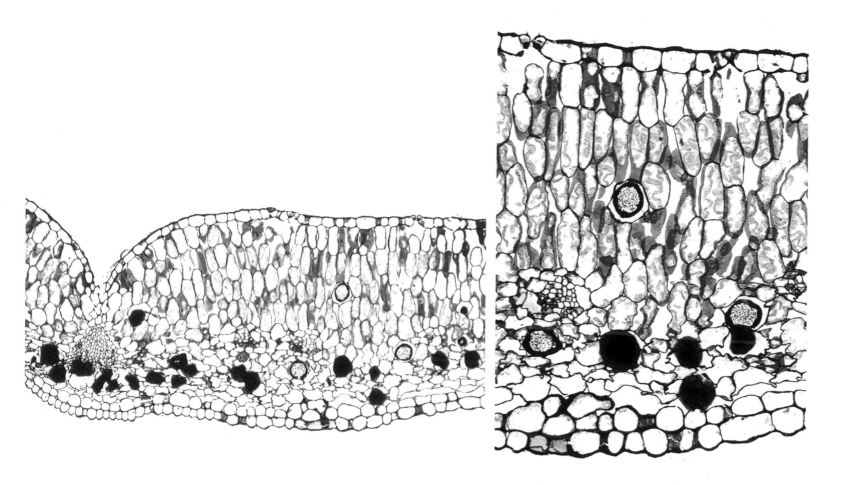

Sections through the leaf of *Epilobium vernicosum*, showing stomates on the upper surface.

These sections through an ***Epilobium vernicosum*** leaf show clearly that the stomates are indeed on the upper surface. Why does this willowherb build its leaves "upside-down"? The bare rocks that it lives in are excellent solar absorbers, and as a result, they are much hotter during the day than even the upperside of the willowherb's leaves exposed to full sunlight. Hence, if the stomates faced *toward* those hot rocks, they would lose more water vapour than if they faced directly into the sun. By wiping stomates altogether from the underside of its leaves, the willowherb markedly increases its chances of survival in its rocky habitat.

44

A showy buttercup, *Ranunculus insignis* (of the **buttercup** family, the **Ranunculaceae**), growing in the sink-holes of Mt. Arthur.

Buttercups live among rocks, too – this is ***Ranunculus insignis*** growing on the walls of a sink-hole in limestone country. Although buttercups thrive among rocks, they do not grow out in the open as the willowherbs do, so their desiccation problem from hot rocks is less severe. Accordingly, their solution to the problem is more subtle than the willowherbs'– they have stomates on *both* their leaf surfaces. They start out the day as other plants do, with their lower stomates open. But as the rocks under them warm up, they risk drying out. Automatic controls in their leaves gradually close down the lower stomates and open the upper ones instead. They carry on that way until late in the day when the temperature of their upper leaf surface drops below the temperature of their lower leaf surface.

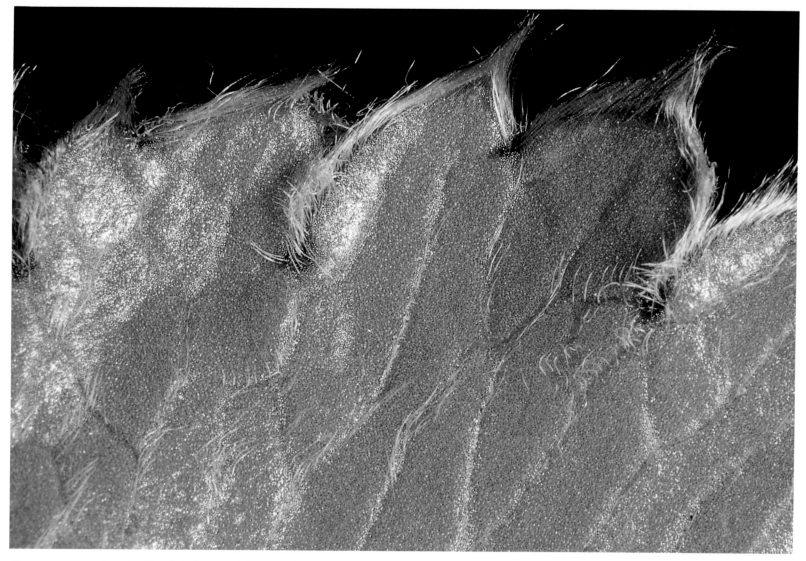

The upperside and underside of the leaves of *Ranunculus insignis*.

Ranunculus insignis has picked up that automatic machinery during the thousands of years it has been evolving. And how did it do that? Any natural population has "genetic odd-balls" in it, and long ago the **Ranunculus insignis** odd-balls that by chance built stomates on both their leaf surfaces grew bigger and lived longer than their less inventive cobbers, and so they left more offspring behind them. Their offspring inevitably did the same in their turn, and that happy outcome carried on for generation after generation. By consistently siring a larger proportion than usual of the next generation, their talent for avoiding drying out eventually spread right through the buttercup population, and so by now they all can do it.

Two snow grasses growing on Mt. Arthur, the pale-ribbed snow grass, *Chionochloa pallens*, and the carpet grass, *Chionochloa australis* (both of the **grass** family, the **Gramineae**).

The snow grasses dominate New Zealand's alpine zone. On the left is a variety of the pale-ribbed snow grass ***Chionochloa pallens***, and on the right in the foreground is the carpet grass ***Chionochloa australis***. Carpet grass grows in the mountains of the South Island from Nelson to Arthurs Pass, forming large patches of sward so dense that it excludes most other plants. Its welcoming common name of carpet grass is deceptive – it can be painful to sit down on, because its leaves have sharp points. The leaves are slippery, too – unwary trampers have suffered some memorable bruises sidling across steep slopes covered with carpet grass, and an unlucky few have fallen to serious injury and death.

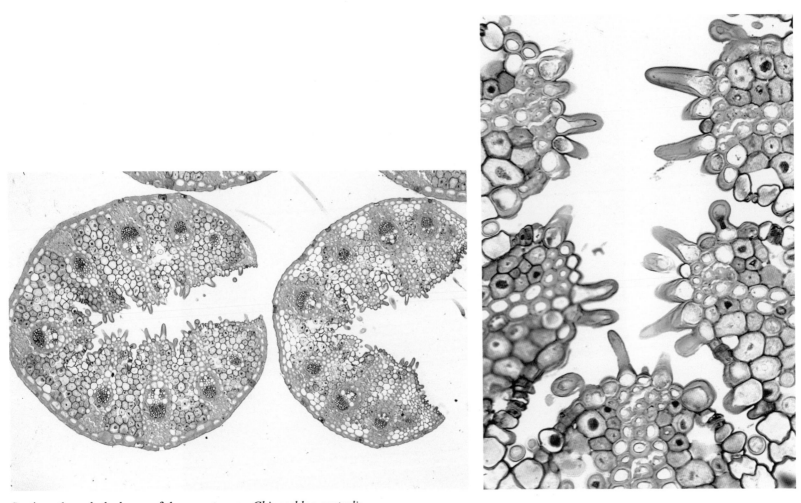

Sections through the leaves of the carpet grass, *Chionochloa australis.*

With their long thin leaves, the snow grasses are more at risk of drying out in strong winds than are most other alpine plants. Not surprisingly, they have some highly effective automatic machinery for conserving their water and so avoiding desiccation. That water-holding machinery is part of their leaf structure. These leaf-sections show how the carpet grass conserves its water. The leaf is shaped like the letter **c** with a hinge in its middle. All the stomates are inside the **c**, which closes up tightly when the leaf wilts, preventing the loss of water vapour through the stomates. But even more cunning, the **c** need not be completely shut to begin conserving water because the stomates are in hair-lined grooves which collapse when the leaf first begins to wilt – as the grooves collapse, the hairs lining them push together and interlock, and the closer the hairs get, the more they interfere with air movement next to the stomates. As a result, the carpet grass is able to respond sensitively to even slight changes in its water supply.

Chionochloa pallens on the flanks of Mt. Arthur.

The leaves of **Chionochloa pallens** are flat rather than circular like **Chionochloa australis'** leaves, but they conserve their water in much the same way – their stomates too are in hair-lined grooves, and as their leaves start to wilt, the grooves collapse, spindling the leaves into tight cylinders filled with dead-still air.

The snow grasses pay a price for the way they conserve water. With only still air next to their stomates, they have trouble pulling in enough carbon dioxide, which is a raw material of their photosynthesis (they turn the energy-poor carbon dioxide into energy-rich carbohydrates, making up the energy difference with the sunlight trapped by their chlorophyll). As a result, snow grasses can grow only slowly. They survive, though, because they can photosynthesize flat-out at temperatures barely above freezing, they demand little energy for their day-to-day running, and they make their leaves work for more than just one season.

trapping enough sunlight

The snow marguerite, *Dolichoglottis scorzoneroides* (of the **daisy** family, the **Compositae**), our largest and showiest mountain daisy.

We have seen how alpine plants solve problems of capturing mineral nutrients and conserving water, but what about energy? Alpine areas often are clouded over, and you might think that the plants growing there could suffer a serious shortage of sunlight. Some do, but not from thick cloud, as it turns out – rather, their worst threat comes from other plants shading them out, and they have come up with a variety of successful defences against that threat.

Hebe topiaria (of the **figwort** family, the **Scrophulariaceae**).

This hebe is named ***topiaria*** because it looks like a topiary, a shrub that has been trimmed into an artificial shape. What does such an unlikely shape do for this hebe? Besides spreading the hebe's leaves in all directions equally so that they capture a maximum of sunlight, it suppresses the hebe's **competitors** by shading them out. Said another way, this hebe's shape is a solution to the problem of competition for sunlight.

All living things suffer the stress of competition, and they can respond to that stress in three ways, ways familiar to us because we use them when we try to solve our own competition problems. For example, if we face competition for a job, we can (1) just grudgingly put up with the competition, (2) we can try to win the competition (but of course we risk losing), and (3) we can change ourselves enough that the competition simply disappears (by learning another skill or by shifting to another town). The first and third strategies do not work with plants competing for sunlight - after all, green plants *must* have sunlight. That leaves trying to win, and this hebe's odd shape does just that. It corners a private supply of sunlight for the hebe by getting between the sun and any competing plants – it overtops them, shades them out.

55

A portion of a twig of a snowberry, *Gaultheria crassa* (of the **heath** family, the **Ericaceae**).

Shrubs and trees can overtop quickly and efficiently because they grow by what an engineer would call "modular construction" – they produce just a single building unit (a leaf plus a portion of stem plus a bud) and they grow by making endless copies of that unit, sticking them together end-to-end, as in this snippet of a snowberry twig.

The porcupine shrub, *Melicytus alpinus* (of the **violet** family, the **Violaceae**), and a crustose lichen *Haematomma babingtonii* growing on its stems.

Even with their talent for modular growth, shrubs and trees nonetheless spend a lot of energy and materials to win the competition for sunlight – much of the sunlight that they corner for themselves by growing tall goes into building their trunks, branches, and myriad twigs. Not surprisingly, other alpine plants have found ways to exploit that heavy expenditure to their own advantage. For example, this lichen gets into the sun on the cheap by growing on the stems of alpine shrubs – that piggy-back life-style makes it an epiphyte, which means "on a plant". Its obliging host here is the woody alpine violet *Melicytus alpinus*. Because the lichen does not harm its host, it goes by the name of **commensal** rather than parasite.

A forstera, *Forstera tenella* (of the **stylidium** family, the **Stylidiaceae**).

Many herbaceous alpine plants solve the problem of over-topping by growing tall themselves. This forstera sends up a tall stem that does triple duty – it gets the flowers into full view of pollinating insects, raises the leaves well off the ground and into the sun, and traps sunlight from all angles (including sky radiation, the blue light of the sky). The stem is photosynthetic, even though its colour is red rather than green (the green chlorophyll is there, but is masked by a dark red pigment called anthocyanin, which protects the plant by absorbing damaging short-wavelength ultraviolet radiation).

Tall herbs like forsteras have solved the problem of competition for sunlight more cheaply and more safely than tall shrubs have – more cheaply because they do not spend the energy and materials to build stout woody tissue, and more safely because they die back to a cluster of hardy leaves or else they overwinter as dormant seeds, while shrubs are exposed to drying winds and damaging ice-blast all winter long. In fact, tall shrubs die out with altitude because their summer growth can not compensate for that winter die-back. Only so-called prostrate shrubs can survive at high altitudes. The low growth of prostrate shrubs is genetically controlled, but all plants are stunted by wind and other touch (thigmic) stresses – the constant bending triggers short and stocky growth. That response explains the gnarled look of trees at timberline and also the spindly look of plants grown in wind-free glasshouses.

A cushion plant, *Phyllachne colensoi* (of the **stylidium** family, the **Stylidiaceae**).

The growth form called a cushion solves together the problems of winter ice-blast and over-topping, plus a third problem of grazing – a cushion is low enough to escape winter winds, dense enough that other plants can not get a foot in, and hard and slippery enough that grazers can not bite into it.

Phyllachne is thought to be among the oldest New Zealand flowering plants, a truly ancient relict from 60 million years ago when the country was largely flat and covered in evergreen forest. Probably **Phyllachne** at that time lived on infertile soils in forest openings. When the mountains rose up some 6 million years ago, **Phyllachne** rose with them, still living on infertile soils but facing the added hazards of cold winds and frequent freeze-thaw cycles. By good chance, its cushion growth form allowed it to cope with those hazards, serving it well during the glacial periods that drove countless other plants into extinction.

61

An unnamed cushion-forming species of *Kelleria* (formerly *Drapetes*, of the **daphne** family, the **Thymelaeaceae**).

Like other cushions, this one is made up of a dense mat of branches running along the ground and sending up closely packed leafy shoots – on the right is a single branch teased out from the cushion and photographed from the side. Each branch has so few leaves that it can not trap much sunlight – an ecologist would say that it has low **productivity**. However, the cushion packs so many branches tightly together that as a whole it easily matches the productivity of much bigger plants like mountain daisies.

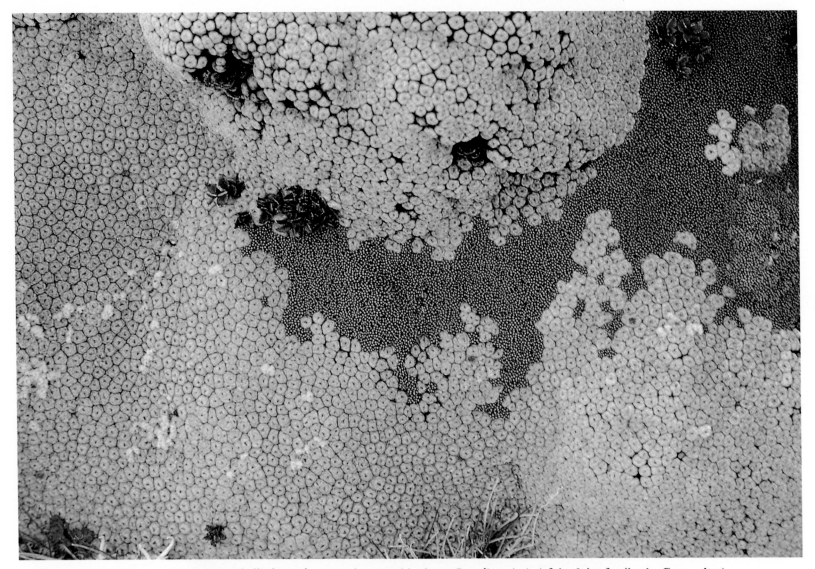

A seamless join between a cushion plant, *Phyllachne colensoi*, and a vegetable sheep, *Raoulia eximia* (of the **daisy** family, the **Compositae**).

Cushions are made by a fully automatic mechanism that stops their branches from growing any higher than whatever is next to them. That precise control of their growth allows them to make seamless joins with rocks and with other cushion plants, thus minimizing damage by predators or winter ice-blast, and invasion by competitors that could shade them out.

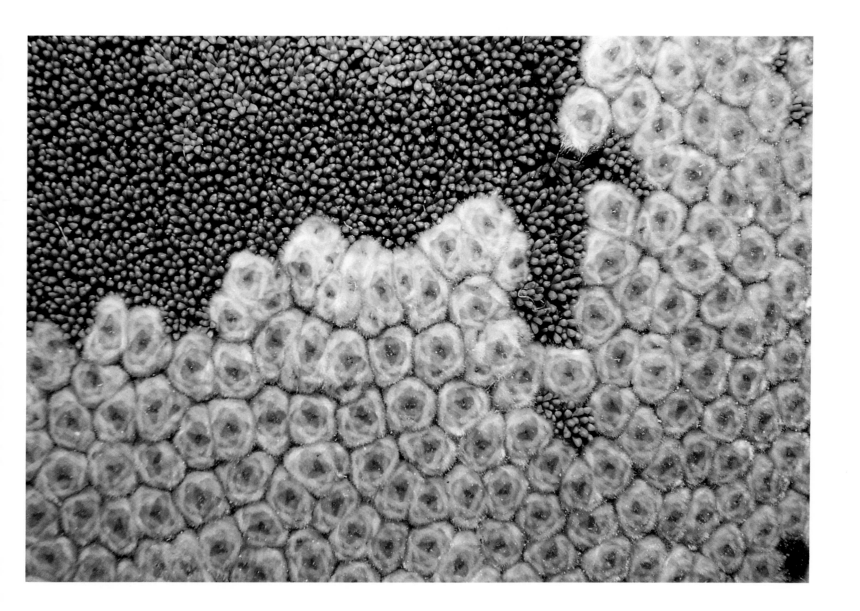

One of New Zealand's famed vegetable sheep, *Raoulia eximia* (of the **daisy** family, the **Compositae**).

Botanists from all over the world travel to New Zealand to see vegetable sheep like this ***Raoulia eximia***. New Zealand boasts far more cushion plants than similar alpine areas in the northern hemisphere, probably because of our milder winters and longer snow-free growing season (our alpine areas have a low average temperature, but they do not suffer from great extremes of cold). Those mild conditions also allow our cushions to get by without overwintering buds or underground food reserves.

On the left ***Raoulia eximia*** is growing with several species of ***Celmisia*** mountain daisies as well as a ***Phyllachne colensoi*** cushion. In the close-up on the right are its heads of dark red flowers, and some of its dandelion-like fruits that are typical of members of the daisy family.

fending off the bugs and beasts

The giant vegetable sheep, *Haastia pulvinaris* (of the **daisy** family, the **Compositae**).

New Zealand's most spectacular cushion-former is ***Haastia pulvinaris***, here growing on a stable rock-field. Besides fending off grazing predators, squeezing out competitors, and standing up to winter ice-blast, such large cushions are protection against drought, because water in the soil beneath them can not evaporate through the tangle of waterproof hairs on both surfaces of their densely packed leaves.

Hare droppings and a dwarf eyebright, *Euphrasia zelandica* (of the **figwort** family, the **Scrophulariaceae**).

Given enough time, the vegetation of any area with a settled climate eventually stabilizes, but New Zealand's mountains face a restless future now that modern man has introduced European grazers. Hare droppings like these are a common sight in subalpine areas (the tiny plant growing among them is a root-parasitic eyebright called *Euphrasia zelandica*). Our alpine and subalpine plant communities nowadays change constantly under the grazing stress of hares, deer, and goats.

The early rapid spread of introduced grazers heavily damaged herbfields and tall grasslands. Eventually the grazers were brought under control in large areas by government-paid cullers and by shooters supplying the buoyant overseas markets for wild venison and goatmeat. Deer and goats were nearly finished off in some places by helicopters meeting the booming demand for captured feral farm stock. As a result, the herbfields and grasslands recovered dramatically from their earlier collapse. But now deer- and goat-farmers are breeding their own stock more cheaply, all but ending live-capture by helicopter, and so the wild grazers are again increasing rapidly, eating back the most palatable alpine species to inaccessible rocky outcrops and high altitudes.

Such grazing disturbance in tussock areas has already allowed foreign plants to get established. Long before the massive disturbance of modern times, seeds were being carried in from all over the world (especially from Australia), but they could not survive here because they could not displace native species. However, now that man has disturbed native areas, they *can* survive, and are forming new plant communities.

73

Two well-camouflaged plants – *Parahebe cheesemanii* (of the **figwort** family, the **Scrophulariaceae**), and a haastia, *Haastia sinclairii* (of the **daisy** family, the **Compositae**).

Being rooted to the ground, plants can not run away from their predators, but they can at least hide, and some are masters of camouflage. This parahebe is nearly invisible when not in flower, and the haastia blends into its rocky home whether in flower or not. Of course, these plants do not somehow take note of the colour of their background and then cleverly camouflage themselves – rather, they long ago came in a wide variety of colours, and over the years the poorest matches were picked off by hungry predators, leaving the best camouflaged to parent future generations. By now, each local population uncannily matches the unique colour and texture of its habitat.

A forget-me-not, *Myosotis angustata* (of the **borage** family, the **Boraginaceae**).

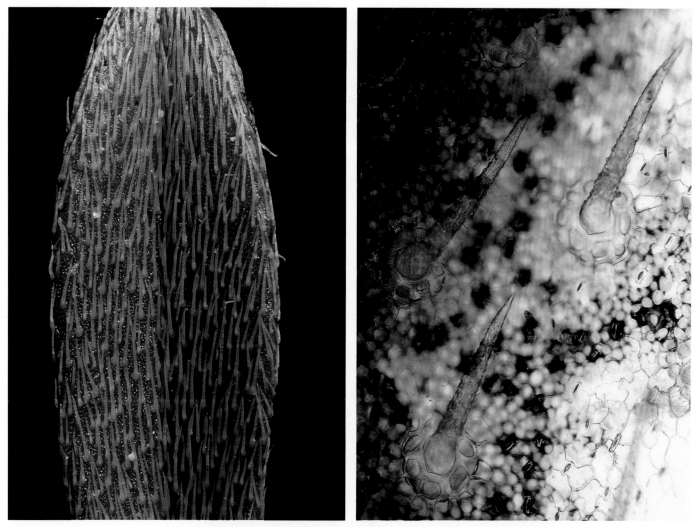

Close-up and microscope views of the leaf hairs of *Myosotis angustata*.

Cushions and camouflage are only two of the defences that our alpine plants use against their predators – hairs like these on the leaves and stems of forget-me-nots fend off hungry insects trying to tuck into the plants for din-dins. *Myosotis* was named by the famous 18th Century Swedish botanist Linnaeus, the popularizer of the Latin binomial system of naming plants – he chose the name ***Myosotis*** for the forget-me-nots because their hairy leaves reminded him of a mouse's ear (in classic Greek, ***Myos*** means "mouse" and ***otis*** means "ear"). Hairs defend plants from leaf-eating insects in several ways. For a start, an insect can not get a good purchase on the leaf while stumbling through the hairs. When it slips, it must spend energy catching itself and gaining another precarious foothold. It soon tires and looks for its calories elsewhere. Also, many plants fill the interior of their hairs with noxious irritating and poisonous chemicals. An insect eating or just breaking off the hairs gets a dose of a repellent. The tough walls of the hairs are a defence, too – they wear down the insect's mandibles, so it ends up eating less, and perhaps can not mature before winter sets in. Even if it does manage to mature, it moults or pupates into a smaller adult, and so lays fewer eggs during the next season.

An alpine bidi-bidi, *Acaena anserinifolia* (of the **rose** family, the **Rosaceae**).

No tramper has escaped the pesky barbed fruits of bidi-bidis hitching a free ride on socks and trousers. The leaves of this common but highly variable bidi-bidi are often downy with hairs that fend off leaf-eating predators. Hairs like these are structures that allow the bidi-bidis and other alpine plants to keep the vital supplies of energy and materials that they have managed to capture from their environment.

77

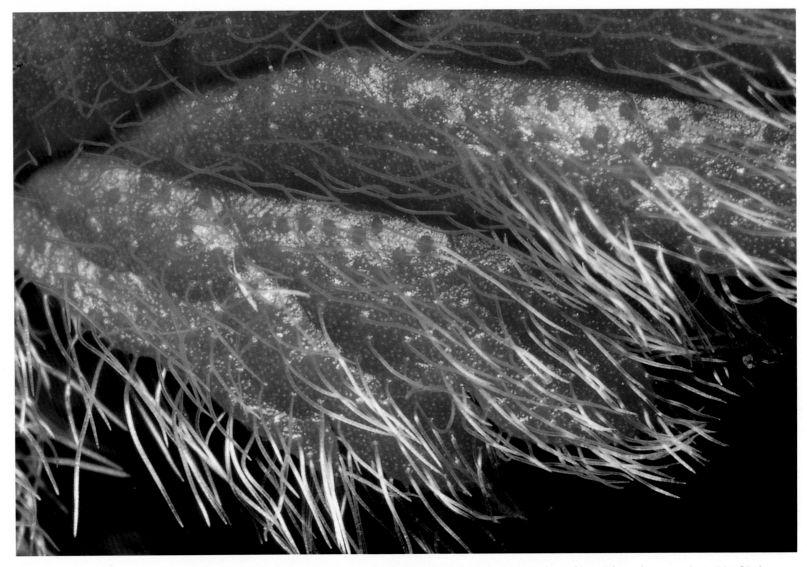

The leaves of an alpine avens, *Geum leiospermum* (of the **rose** family, the **Rosaceae**), and a microscope view of its anti-grazing crystals and leaf-hairs photographed under polarizing optics.

Like the downy hairs of the bidi-bidis, the stout hairs on the leaves of this alpine avens are a defence against leaf-eaters. External defences like hairs can be highly effective, but many alpine plants have turned to internal defences as well, and the alpine avens and bidi-bidis are among them – crystals of an insect-deterring chemical lie just under the surface of their leaves. The crystals irritate the mouth-parts of leaf-eating insects, and they are so dense inside the leaf that an insect can not take more than a few bites without chomping into one. You see hairs as well as crystals in this photograph of a leaf of an alpine avens that was firmly squashed and then viewed under polarized light.

79

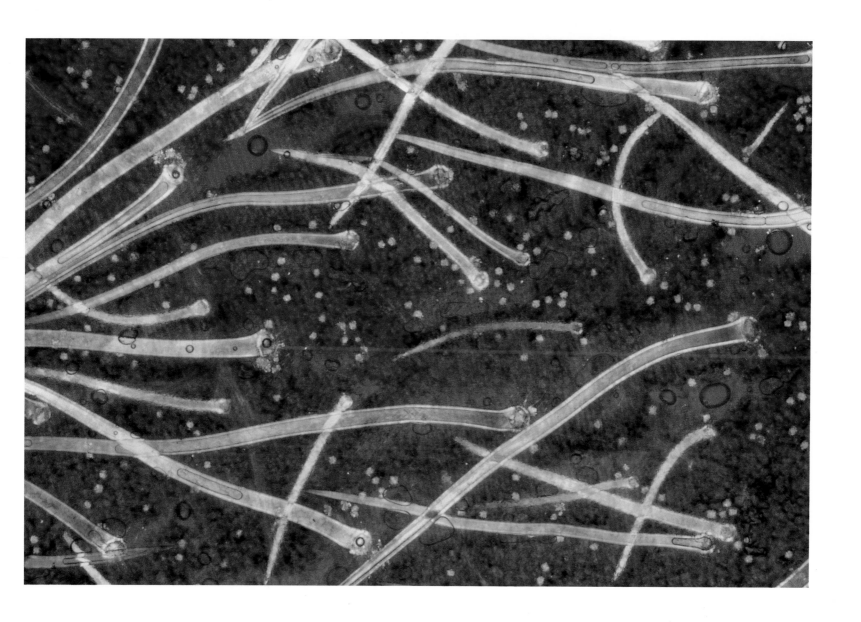

A native oxalis, *Oxalis magellanica* (of the **wood-sorrel** family, the **Oxalidaceae**).

Many of New Zealand's alpine plants lace their leaves, stems, and even flowers with insect-deterring chemicals, but the best examples are the southern heaths (the epacrids like ***Epacris, Dracophyllum, Leucopogon,*** and ***Pentachondra***), the willowherbs (***Epilobium***), and the wood-sorrels (***Oxalis***). The leaves of this dainty native ***Oxalis magellanica*** resemble the all-too-familiar introduced oxalis that gardeners have long cursed as an aggressive weed almost impossible to eradicate. ***Oxalis*** often is eaten in salads for its pleasantly acid flavour (***oxalis*** in Greek means "sour"). That tangy flavour comes from oxalic acid.

82

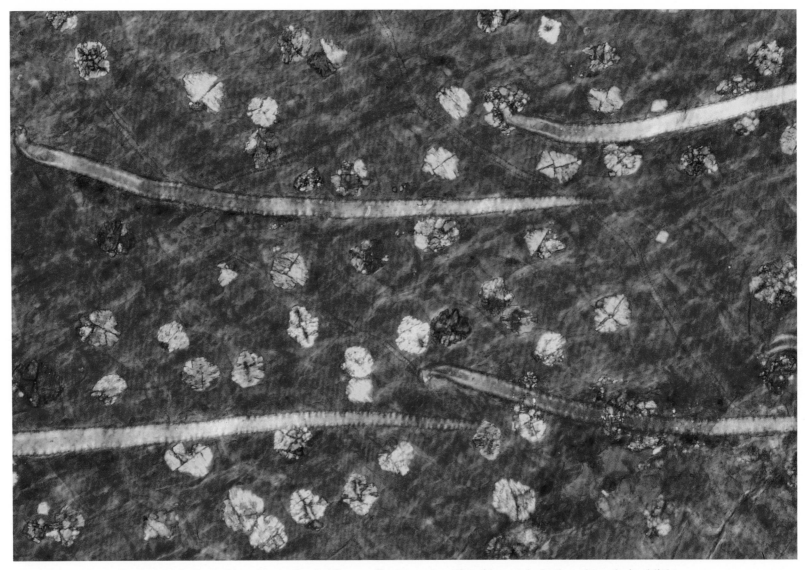

Microscope views of a squash of a leaf of a native oxalis, *Oxalis magellanica,* and oxalic acid crystals, both under polarized light.

The oxalate in **Oxalis** poisons grazing animals by damaging their kidney tubules and by tying up vital soluble calcium as non-soluble calcium oxalate, causing hypocalcaemia (commonly called "milk fever") with its bizarre symptoms of staggering and convulsions. Oxalic acid is not likely to harm people in only salad quantities, but it does kill people who drink wines illegally sweetened by ethylene glycol, a cheap anti-freeze. The ethylene glycol is oxidized to the toxic oxalic acid by a natural body enzyme. The cure is a drunkard's dream, a nearly intoxicating dose of pure alcohol – the alcohol competes with the ethylene glycol for the attention of the enzyme, and so while the enzyme is tied up working flat-out on the alcohol, the ethylene glycol is quietly excreted before it can be oxidized to poisonous oxalic acid.

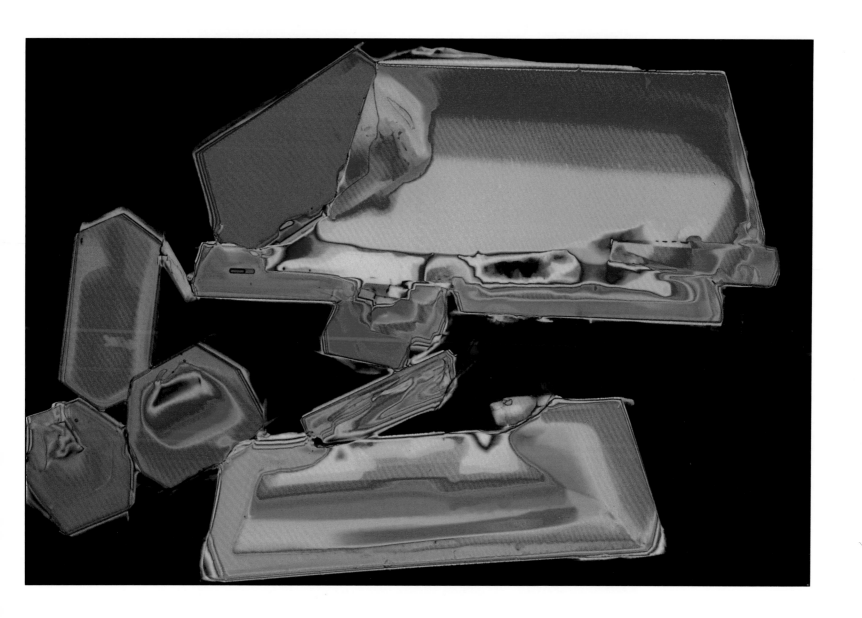

85

Two epacrids (of the **epacrid** family, the **Epacridaceae**) – a grass tree or turpentine bush, *Dracophyllum longifolium*, and a southern heath, *Epacris pauciflora*.

All of New Zealand's epacrids have anti-grazing chemicals in their leaves. On the left is ***Dracophyllum longifolium***, commonly called the grass tree because of its long slender leaves. Those leaves are shot through with chemicals that are irritating, foul-tasting, and poisonous to leaf-eaters. On the right is ***Epacris pauciflora***, a southern heath that is common on the grazed, burnt, and nutrient-poor **pakihi** areas of Golden Bay and the West Coast. Both of these epacrids place their anti-grazing crystals for maximum effect, next to the veins in their leaves. The veins are the internal piping and the strengthening beams of the leaves, and if an insect bit into one, a leaf would lose not only precious water but also its strength to resist alpine winds.

A rock-loving willowherb, *Epilobium glabellum* (of the **evening-primrose** family, the **Onagraceae**), and a microscope view of the sharp needle-like raphide crystals inside its leaves.

Some of our alpine willowherbs live out in the open on bare rock, an easy target for alpine browsers like hares, goats, and deer, so it should not surprise us that willowherbs have a strategy to keep those predators at bay. The strategy that they use is common in plants – they systematically place thousands of sharp needles of insoluble calcium oxalate in their leaf tissues. The needles (called raphides) punch myriad tiny holes in the soft lining of the mouths and stomachs of grazers. That pin-cushion strategy would not be anything more than a nuisance to grazers, except that plants which produce raphide crystals produce as well a formidable array of powerful irritants and poisons like protein-dissolving enzymes. The enzymes trigger the release of histamines that swell the grazer's mouth and cause intense pain.

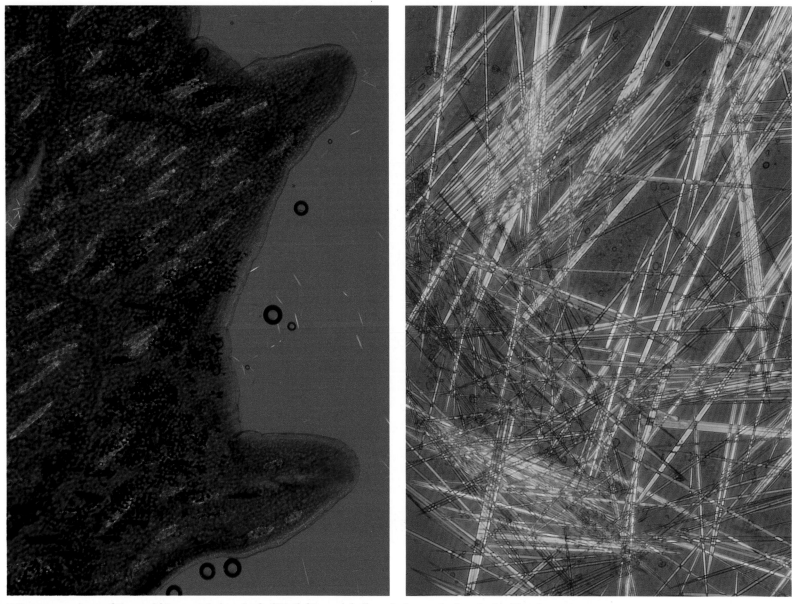

Microscope views of the raphide crystals in a leaf of *Epilobium glabellum*, before and after the leaf has been chewed.

The willowherbs produce their sharp raphides in neat bundles lined up inside their leaves. On the left is a microscope view of the margin of a glossy willowherb leaf – the yellow blobs are the bundles of needles. The bundles break up quickly as the leaf is chewed, and the needles scatter into what looks like the children's game of "pick-up-sticks". The bundles of raphides are so dense inside the leaf that a predator even just cautiously sampling the leaf is soon put off.

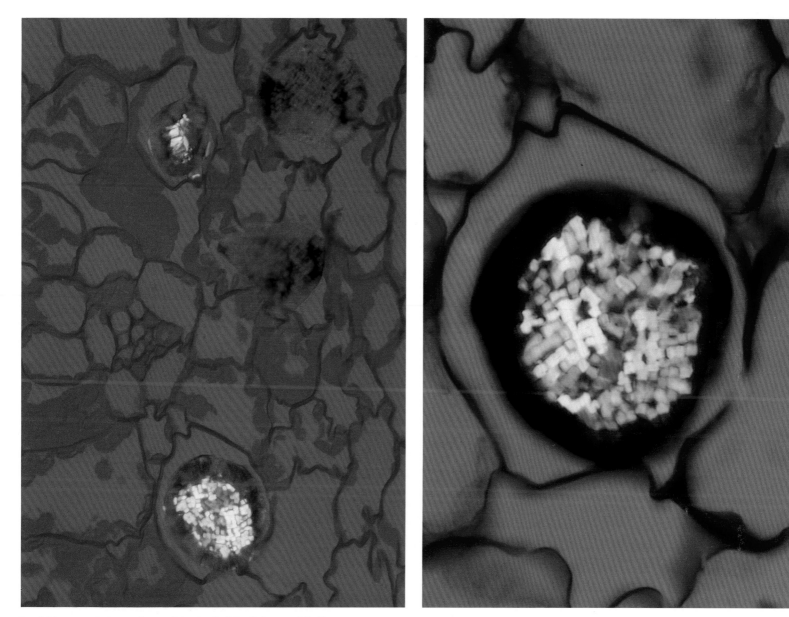

Raphide crystals in sections of the leaf of *Epilobium glabellum*.

The bundles of raphides show up well in sections of leaves, too – here you see bundles of the needles cut cross-wise. Each bundle contains about 200 needles. The densely staining capsules around the bundles may contain an irritating substance that invades the thousands of holes left by the sharp needles in the grazer's mouth lining.

A native alpine St. Johnswort, *Hypericum japonicum* (of the **St. Johnswort** family, the **Hypericaceae**).

This St. Johnswort goes right for the jugulars of its predators – in the tiny black dots along the margins of its leaves and flower petals, it produces a pigment called hypericin that lodges in the skin of any animals that graze it. If that skin is not already pigmented, the hypericin absorbs sunlight, and passes on the energy to the animal's tissues, causing sores and ulcers, blindness, convulsions, and even death. Thousands of sheep in South Island high-country have fallen victim to St. Johnsworts. After grazing the plants, the sheep are ultra-sensitive to contact with cold water, and when being driven across rivers or streams, they rear up and thrash about. If fully submerged (as in the compulsory yearly de-lousing dip), they convulse violently and can drown.

gifting solutions to the next generation

A mountain portulaca, *Montia calycina* (of the **portulaca** family, the **Portulacaceae**).

This alpine portulaca and all the other plants that survive today in our mountains are the descendants of plants that stumbled upon successful solutions to their survival problems during the thousands of years of their evolution, problems like finding enough mineral nutrients and water, trapping enough sunlight, and fending off sustained attacks from their predators and competitors. Each new solution allowed its lucky owner to grow bigger and to live longer and to leave more offspring behind it. Its offspring did the same in their turn, and so did theirs. By thus parenting an ever larger percentage of succeeding generations, its descendants eventually gifted the solution of the problem to the entire species. That steady process is called evolution by natural selection.

93

A whipcord hebe, *Hebe ochracea* (of the **figwort** family, the **Scrophulariaceae**).

If it is of value for alpine plants to *find* solutions to their survival problems, then it is of equal value for them to *keep* those solutions, and **breeding barriers** do that job for them – breeding barriers wall off their successful genes from dilution by ordinary "garden-variety" genes. The building of such breeding barriers inevitably creates new species, which is why there are so many different kinds of creatures in the world. *A species is a natural population that has solved survival problems and also protected its solutions with breeding barriers.*

A hebe with both white and purple flowers. This species and *Hebe ochracea* are largely restricted to the Nelson province.

However, the breeding barriers around a species always are *leaky*, and genes enter and leave the species' **gene pool** through the leaks, altering the species and occasionally creating new species. The leaks come in all sizes, some so massive that they are easier to find than the barriers themselves are. Other leaks are so tiny that they show up only rarely – the many species of willowherbs in New Zealand are a result of such minute leaks over several million years. But in all cases a barrier and its leaks *together* are a **system** that controls a species' genetic isolation. This handsome hebe usually has white flowers, but it produces mutant purple flowers often enough to catch the attention of horticulturists who are keen to breed garden ornamentals with showy coloured flowers.

The snow totara, *Podocarpus nivalis* (of the **podocarp** family, the **Podocarpaceae**) – on the left is a fleshy female **aril** with its single surviving seed, and on the right are male **cones** spilling their pollen.

Control systems work by pitting one thing against another. For example, the speed of a car is controlled by pitting the car's brakes against the accelerator. By analogy, breeding barriers are the "brakes" of the system that controls a species genetic isolation – they *prevent* gene exchange (and so are called **isolating mechanisms**). Pitted against the barriers are the "accelerators" of the system, the leaks – they *permit* gene exchange. Hence breeding barriers and their leaks are the two opposing halves of a control system that not only protects a species' existing solutions to survival problems, but also permits experiments with novel untried solutions.

A shrub groundsel, *Brachyglottis adamsii* (of the **daisy** family, the **Compositae**), in splendid flower in the Cobb Valley.

The species' environment determines whether or not a survival solution is successful. A solution *is* successful if it allows the species to leave more offspring in that environment (the species then is said to be **fitter**, and the solution is said to be **adaptive)**. Eventually, adaptive genetic information dominates the species' gene pool. The unlucky owners of unsuccessful solutions linger on producing fewer offspring, or else they peg out, and so eventually their maladaptive genetic information drops to a low level in the species' gene pool or even disappears altogether.

One of New Zealand's showier native orchids, *Caladenia lyallii* (of the **orchid** family, the **Orchidaceae**).

That process is called **natural selection**, and it is the **feedback link** in the fully automatic system that controls a species' genetic isolation. If existing solutions to survival problems are working well, then novel solutions do not do the job any better, and they just die out. Conversely, if existing solutions are *not* coping well (perhaps because the environment has changed), then novel solutions have a better chance, and their lucky owners automatically parent more offspring. Of course, if *no* novel solutions turn up to cope with pressing survival problems, the species slides into extinction. Lots have, and in fact over the long haul of geologic time, far more species have gone extinct than survive today.

103

Stamens spilling their pollen, and a stigma ready to receive pollen – genetic information is "scrambled" during the process that produces sperm and egg nuclei in plants, and then during pollination that genetic information is exchanged among the plants of each species.

What are breeding barriers and their leaks actually like? They come in many forms, but virtually all plant species avoid crossing with plants unrelated to them by interfering with the germination of foreign **pollen** on their **stigmas**. The few pollen grains that do manage to germinate suffer a sustained chemical attack as their **pollen tubes** grow down through the flower's **style**, and that prevents the setting of any viable seed.

The simplest yet most effective breeding barrier just divides up a species' genetic information among **chromosomes** that are unique in shape and size, and so can not match up with the chromosomes of other species growing nearby, hence foiling any chance of a fertile cross. For a cross to be fertile (to produce offspring), all of the many chromosomes must match up during **meiosis**, the division of cell nuclei that produces sperm and egg nuclei during plant reproduction.

A snow-berry, *Pernettya macrostigma* (of the **heath** family, the **Ericaceae**).

Some breeding barriers are geographical – two species that live far apart are able to cross, but they never actually do, because they just do not get the chance to. They had a common ancestor long ago, but got separated somehow, perhaps by continental drift or by slowly rising sea levels. They readily cross if they get together again, as often happens in gardens – many new and valuable plant cultivars (**culti**-vated **var**-ieties) have popped up that way to the surprise and delight of gardeners. Sometimes plants belonging to even different genera can cross – for example, our native *Gaultheria* and *Pernettya* snow-berries cross shamelessly, and only expert field botanists and biochemists can figure out which of the several species in the two genera are the parents of the hybrids.

Two closely related gentians, *Gentiana tenuifolia* (left) and an unnamed species from the Cobb Valley (both of the **gentian** family, the **Gentianaceae**).

Some species exploit time as a breeding barrier. For example, most gentians bloom during late summer. Bucking that general rule, the unnamed Cobb Valley gentian on the right blooms during early summer, and most of its flowers are gone by late summer. As a result, its unique genes are not easily diluted by the genes of any other gentians. These two species of gentians differ in their size and habitat as well as their flowering period, yet their flowers often look much alike except for the dark purple pigment on the fuzzy pollen-trapping stigmas of *Gentiana tenuifolia.*

The bristle-tussock, *Rytidosperma setifolium* (of the **grass** family, the **Gramineae**).

Some time-isolated species manage to stay separate with much less time than a season between them – in fact, they make do with only a few hours. The florets of grasses open at a precise time each day, releasing their pollen and exposing their stigmas to the wind, and then they close again smartly. As a result, grass species that *could* cross in fact rarely do, even when living side-by-side, because their florets are open at different times of the day. Such precision breeding barriers spring random leaks when unusual weather disrupts the timing of floret-opening and pollen-release, allowing the mixing of genes.

An alpine avens, *Geum leiospermum* (of the **rose** family, the **Rosaceae**).

A species could stay separate by producing an unusual flower that attracts and "fits" only a single insect pollinator. That would be a handy breeding barrier, but as it happens, New Zealand's alpine plants can not use it, because their mountain home simply does not have enough different kinds of specialized pollinators like long-tongued bees. Any of our alpines that by chance produce flowers with unusual shapes or colours just do not get pollinated, and so they die out. With so few types of pollinators buzzing around, only the **promiscuous** flowers get pollinated, flowers that attract any and all insects. That explains why our alpine flowers look so much alike (some say monotonous) – most are small, white, and built like a shallow cup or a tube, and so they attract all insects on the wing and fit almost any that plunk down looking for a feed. Hover-flies, blow-flies, tachinid flies, native bees, butterflies and moths, and even beetles visit the flowers of most alpine plants. The flower of this alpine avens is small, white, simple, and unspecialized, and it attracts a great variety of insects rather than only a few specific pollinators.

A subalpine pimelea, *Pimelea longifolia* (of the **daphne** family, the **Thymelaeaceae**).

Plants of the daphne family are prized the world over for their fragrant flowers, and our alpine pimeleas persist in that pleasant family tradition. It might seem that scented alpine flowers could not reliably attract pollinating insects, because mountain air is turbulent – the turbulence breaks up the "scent cone" downwind of the flowers and confuses any insects searching upwind for the source. In fact, though, that rarely matters, because insects are more likely to be out searching for nectar sources on calm sunny days anyway. A fragrant flower then has a distinct advantage over its drab scentless rivals, and a pollinator on the wing looking for a feed quickly homes in. Although the days are seldom calm in the mountains, that too rarely matters, because many New Zealand alpine plants flower for long periods (weeks, even up to two months), and single flowers often last for a week.

Female and male flowers of a prostrate coprosma, *Coprosma perpusilla* (of the **madder** family, the **Rubiaceae**).

With a lot of wind in alpine areas, it is not surprising that many alpine plants are wind-pollinated. We expect that of monocots like the grasses, sedges, and rushes, but some dicots like this prostrate coprosma are also wind-pollinated, probably a response to the lack of specialized insect pollinators in New Zealand during the millions of years of its geologic history. Coprosmas are among the most ancient of our native flowering plants, dating from 60 million years ago when the land was flat and warm.

Almost a fifth of New Zealand's flowering plants have single-sex flowers of some sort, and the coprosmas are among them. Separation of the sexes prevents self-pollination, which is a problem for flowers that have both stamens and stigmas. A plant's pollen and stigmas can be separated in many ways (in time as well as space), but the coprosmas have opted for the no-nonsense solution of packing them off into separate plants. The flowers of both sexes of this prostrate coprosma have drab petals, which is typical of flowers that do not attract insects. While most coprosmas have only two style-arms, this one has four, an adaptation that doubles its chances of snatching pollen grains from turbulent mountain air.

Ripe fruits of *Coprosma perpusilla* and the sun-bleached remains of a meal of coprosma fruits.

The abundant bright orange and red fruits of the coprosmas add a splash of exuberant colour to alpine meadows in late summer. Keas and other birds reliably gorge on the berries, and you sometimes can find the bleached remains of their feeds. Most alpine birds are strong fliers, and they scatter the coprosmas widely across the landscape. Fully half of the alpine plants that produce fleshy fruits like these also separate their sexes, perhaps because they consistently get heavier fruit crops that way, and therefore more readily attract birds.

A New Zealand endemic willowherb *Epilobium vernicosum* (of the **evening-primrose** family, the **Onagraceae**).

Ordinarily, promiscuity destroys plant species by triggering off runaway hybridization – after all, the insect pollinators inevitably carry in mostly maladaptive foreign genes that quickly contaminate a species' unique genes. Knowing that, an overseas botanist was puzzled to learn that New Zealand willowherbs do not protect themselves against hybridization – most of our species of willowherbs can be crossed successfully (they are said to be **interfertile**). That is not true of the overseas willowherbs, so he set out to discover how our species manage to remain distinct in spite of their being interfertile. The answer turned out to be simple – each of our species of willowherbs fertilizes itself. It is *able* to cross with other species, but it rarely does, because its own pollen readily fertilizes its flowers, and few specialized pollinators dust its flowers with foreign pollen anyway. Hence, the creation of new willowherb species is encouraged by the utter lack of any barriers to hybridization, but discouraged by self-fertility of the flowers – clearly, hybridization and self-fertility are the accelerator and brakes of a system that controls the formation of new willowherb species.

115

Sink-holes (collapsed underground caverns) in "karst topography" on Mt. Arthur. Such landscapes are formed by ground water percolating through soluble rocks like limestone.

To survive, though, the new species must have new places to live, and New Zealand's mountains provide them aplenty. Geologically young at a mere 6 million years, they were thrust out of an ancient featureless plain by the grinding collision between two of the earth's gigantic crustal plates. Since then they have been eroded by wind, water, and frost, drenched with rain on one side and dried out by fierce winds on the other. Glaciers have carved huge river valleys out of them, then melted and dumped vast piles of spoil. The resulting habitats are remarkably diverse, and new willowherb species have exploited those habitats with adventurous abandon. By now New Zealand boasts a remarkable 37 of the 200 species of willowherbs in the world, and they are so "ecologically distinct" that a botanist can identify them as easily by their habitat as by their appearance.

The willowherbs are not the only plants to have exploited the diverse and rapidly changing habitats in our mountains during the last 6 million years – the New Zealand flora is dominated by a few large genera with most of their species in alpine and subalpine areas (for example, the coprosmas, hebes, astelias, dracophyllums, and celmisias). Some were already here before the mountains rose up (the coprosmas, astelias, and dracophyllums), but the rest were carried in from distant foreign lands by wind, sea, and birds, and like the willowherbs evolved explosively in the mountains.

A mountain pratia, *Pratia macrodon* (of the **lobelia** family, the **Lobeliaceae**).

The willowherbs are unusual in being virtually self-fertile – most alpine plants readily out-cross with other members of their own species, and in fact they encourage out-crossing by actively preventing self-pollination. They usually do that by separating their pollen from their stigmas in time or place. For example, the anthers of this mountain pratia ripen and spill their pollen before the stigmas have opened up – in fact, the flower even uses the still-closed stigmas to push the pollen out of the flower where insects can get at it (in the close-up on the right, you can see the tightly shut stigmas projecting beyond the cylinder of fused grey anthers). The stigmas open only several days later, after most of the pollen has been carried off. The flower is strictly male while it is awash with pollen and while the stigmas are still tightly closed, but it is almost fully female by the time the pollen is largely gone and the stigmas have finally opened up.

A harebell, *Wahlenbergia marginata* (of the **bellflower** family, the **Campanulaceae**).

The flower of this harebell is now strictly female, with its bottle-brush stigmas fully expanded and awaiting pollen from insect visitors. A few days earlier, though, it was strictly male – the stigmas were firmly shut, and the anthers were spilling their pollen for insects to cart off to other harebells. Only a single withered anther remains as tell-tale evidence of its dramatic sex-change. Plants like the harebells that shed their pollen before their stigmas open are common in our alpine areas, among them the gentians, hebes, flaxes, geraniums, St. Johnsworts, and the daisies and their many relations.

119

A southern heath, *Epacris alpina* (of the **epacris** family, the **Epacridaceae**).

Plants that shed their pollen before exposing their stigmas face a problem of what to do with the spent stamens. Their solutions are as diverse as their flowers. A mountain carrot just cuts the stamens off at their bases, and they fall away. Geraniums lop off only the anthers, then bend the naked filaments inwards. Portulacas and some hebes clip off only the anthers, too, but they bend the naked filaments backwards instead. This southern heath leaves the anthers on, but bends the filaments right back over the petals, well out of the way of visiting insects.

Ranunculus foliosus (of the **buttercup** family, the **Ranunculaceae**), and a close-up of a buttercup flower's many carpels with their receptive stigmas surrounded by immature anthers.

In most "time-staggered" flowers, the anthers ripen before the stigmas do, probably because the anthers can quickly be bent back or even shed after doing their job, whereas the stigmas can not – after all, the stigmas are parts of carpels that later develop into the plant's fruits. Nonetheless, some plants like this buttercup do things the hard way and expose their stigmas first. They later face the problem of how to keep their ripening anthers from spilling pollen onto the exposed stigmas – they solve the problem by bending the stigmas out toward insect visitors only until the anthers are nearly ripe, then pulling them back in again, out of reach of both insects and their own anthers.

A mountain daisy restricted to the Nelson region, Dall's daisy, *Celmisia dallii* (of the **daisy** family, the **Compositae**).

Alpine plants risk being battered and dried out by strong winds in their mountainous home, but at least they can put those same winds to good use to scatter themselves across the landscape. Composites like this Dall's daisy are well adapted to wind-dispersal with their dandelion-like fruits. But, whereas the dandelion's delicate parachute-fruits seem eager to fly off, the fruits of mountain daisies are reluctant adventurers – they cling firmly to the spent heads that spawned them. That adaptation ensures that the fruits fly off only a few at a time during several weeks rather than all at once, increasing the chance that at least some of them will strike conditions just right for their germination, and so will successfully gift to the next generation their solutions to survival problems.

glossary

adaptive – Said of any genetic trait that allows its owner to leave more offspring behind it. s̲e̲e̲ fitness.

blue-green algae – Although long classified as algae, evidence from modern biochemistry and electron microscopy shows that the blue-greens more closely resemble bacteria. Both lack membrane-bounded nuclei, plastids, mitochondria, and Golgi apparatus. As a result, many biologists now lump the two together in a Superkingdom that they call the prokaryotes, fundamentally different from all other creatures, the eukaryotes.

breeding barrier – Anything that prevents the flow of genes from one species to another.

carnivorous plant – Any plant that can lure, trap, and digest insects or other small animals. Sundews *(Drosera)*, Venus' fly-catchers *(Dionaea)*, pitcher plants *(Sarracenia)*, and bladderworts *(Utricularia)* are familiar examples. Their remarkable predatory talents allow them to survive in areas that lack vital nutrients, especially nitrogen and phosphorus – their hapless victims unwittingly bring to them the missing nutrients.

cephalodium (plural cephalodia) – A structure containing nitrogen-fixing blue-green algae either inside or on the surface of a lichen. It gets the name of cephalodium because when sectioned for microscope viewing it looks vaguely like an animal's brain.

chromosomes – Rod-shaped structures in a cell's nucleus that store genetic information in the form of DNA. Each chromosome is made up of an enormously long single strand of DNA protected by proteins coiled and super-coiled into a complex shape. All the members of a species have the same chromosome count in their somatic (non-reproductive) nuclei, usually divided into two similar sets. When the nucleus divides by mitosis, the chromosomes are duplicated exactly, and then are portioned out to the daughter nuclei. As a result, the daughter nuclei are identical to each other and to the original nucleus.

commensal – Any creature that somehow benefits from another creature without actually doing it any harm. Most lichens growing on the bark of trees are commensals – they benefit by getting off the ground and into the sunlight, but do not harm their obliging hosts. However, some large epiphytic lichens (like *Usnea*, old-man's-beard lichen) grow luxuriantly enough to shade their hosts.

competition – An interaction between two creatures (or populations) that harms both. The harm shows up as stunted growth, shortened life-spans, and fewer offspring. The competitors clash because something that they both use is scarce, usually energy (food or sunlight), materials (nutrients or water), or elbow room. Competition is arguably the most damaging stress faced by living things, and has profoundly shaped their evolution. Many creatures solve their competition problems by trying to win (they are aggressive or spew out poisons) or else by avoiding the competition (they change their "niches", the sum total of the demands they make on their environment). Natural selection for truly effective solutions to competition problems has shaped the very structures of living things as well as their packaging into species and even the patterns they make on natural landscapes.

epidermis – The outermost layer of leaves, petals, and most other soft structures of plants. Usually the epidermis is not photosynthetic. Often it grows hairs and is covered with a waterproof waxy layer (called a cuticle).

feedback link – A linkage between the output and the input of a control system that "tells" the control system what to do next. For example, a thermostat "knows" when to turn your room heater on or off because its in-built thermometer constantly compares your room's temperature against the temperature that you want the room to be (and have set on the dial of the thermostat). If the two readings differ, the thermostat sees the difference as an "error". If the room is too cold, it turns the heater on, and if the room is too hot, it turns the heater off. In both cases, the input of the control system is the error reading, and the output is tripping the switch of the heater on or off. The feedback link is the temperature of the room.

   Control systems are of two sorts called DC and DA, short for deviation-counteracting and deviation-amplifying. DC controls stabilize things – they iron out the bumps and hollows, hence are said to work by *negative* feedback (examples are thermostats, automatic pilots, and shock absorbers). In contrast, DA controls accentuate rather than iron out the bumps in things, hence are said to work by *positive* feedback (examples are your growth as a foetus, the growth of populations, and the accumulation of capital in a booming economy).

fitness – Darwinian fitness is measured by how many offspring a creature can leave behind it in a given environment. Hence if one creature has more kiddies than another does, by definition it is fitter. The familiar phrase "survival of the fittest" does not mean what it sounds like to modern ears, the ability to win a marathon or some other test of *athletic* fitness. The history of deer-shooting by helicopter nicely contrasts the two meanings. When deerstalkers first began to use helicopters, deer bolted into the open in a fine display of their athletic fitness, only to be shot dead. The few deer that quietly hid in the bush instead of madly bolting survived to sire the next generation in a fine display of their Darwinian fitness. Non-athletic deer clearly are fitter (in the Darwinian sense) in an environment with helicopter-shooters on the prowl.

gene pool – The sum total of genes in a population at any one time.

haustorium (plural **haustoria**) – A structural connection between the root of a root-parasite and the root of its host. Each haustorium is a direct pipe-to-pipe connection between parasite and host, and so it gives the parasite access to anything coursing through the innermost pipes of its victim.

interfertile – Said of higher plants that are able to cross successfully by producing viable seed.

isolating mechanism – Any trait of a creature that prevents a successful crossing with a member of another population.

meiosis – Cell nuclei can divide by meiosis or mitosis. Meiosis produces daughter nuclei that are genetically different from each other and from the original nucleus, and with only half the chromosome count of the original (meiosis produces the sperm and egg nuclei of most creatures). In contrast, mitosis produces daughter nuclei that are identical to each other and to the original.

natural selection – The many members of a population differ genetically, the inevitable result of mutation and the "scrambling" of genes during meiosis. A lucky few of those genetic variants find themselves the owners of truly effective solutions to their survival problems. Those solutions allow them to grow bigger or to live longer, and so to leave more offspring behind them. Their descendants inherit their good luck in full measure, and so parent an ever larger percentage of succeeding generations. Eventually, the population's gene pool is chock-a-block with genetic traits that solve survival problems. The traits appear to have been selected by an unseen hand, but in fact the selection process is strictly natural and fully automatic.

nitrogen-fixer – Any creature which is able to convert chemically inert nitrogen gas from the atmosphere into the more reactive ammonia, which can then be built into amino acids and proteins. Of all the world's creatures, only some soil bacteria and blue-green algae have the enzymes necessary to fix nitrogen. Legumes somehow encourage nitrogen-fixing *Rhizobium* bacteria to invade their roots and reliably provide them with a private supply of nitrogen. Similarly, many lichens harbour blue-green algae, as do some liverworts, mosses and ferns. Agricultural scientists are keen to isolate nitrogen-fixing genes and splice them into valuable cereal crops like wheat and maize.

pakihi – A Maori word for the nutrient-poor and often water-logged areas of Golden Bay and the West Coast that have been heavily disturbed by fire and grazing.

peat – Partially decomposed dead vegetation. Living things are of two sorts, producers that get their energy by trapping sunlight, and consumers that get their energy by eating the producers. Ordinarily, all producers are gobbled up either alive (by herbivores) or dead (by decomposers), but during ancient geologic periods, there were not enough consumers around to eat all the producers. As a result, their corpses piled up. That happens even nowadays to the sphagnum mosses growing in the sodden soils around alpine tarns, where harsh conditions like high acidity, numbing cold, and low oxygen conspire to nobble the consumers that we call rot-bacteria and fungi. Not only sphagnum mosses are preserved – you may have seen on the telly recently a human corpse dug out of a peat-bog in Britain (he was given the whimsical press name of "Pete Marsh", but is more formally known now as "Tollund Man"). He was dated by radio-carbon analysis at over 2000 years old, yet he was preserved well enough in the peat for forensic scientists to prove that he had come to a sticky end, apparently the victim of a ritual murder.

  Usually producers are at least partly eaten when they die, so that after a while they do not look like they were ever alive – they turn black or gooey and are called organic sediments. Peat and muck soils are deposits of such partly-eaten producers. If buried in bulk by sediments and then heated and compressed, they change into coal, oil, and gas.

photosynthesis – A complex system of linked chemical reactions that enriches an energy-poor raw material (usually carbon dioxide) using the energy of sunlight. The energy of the sunlight is absorbed by a pigment (usually chlorophyll), and is used to split water molecules (the other raw material of photosynthesis), releasing electrons and producing oxygen as a waste product. Most of the electrons are then added to carbon dioxide, producing an energy-rich carbohydrate. The plant burns some of that carbohydrate as a fuel (just as we harvest the energy of the foods that we eat). However, the plant uses most of the energy-rich carbohydrate as "carbon-skeletons" to build the myriad organic molecules that make up its structure and run its day-to day metabolism.

pollen – Massed pollen grains. Pollen grains are produced in flowering plants inside the sac-like anthers at the tips of a flower's stamens. A pollen grain is one of two kinds of spore produced by higher plants. The other kind remains hidden inside the plant, and can be seen only under a microscope after sectioning and staining the plant.

pollen tube – If a pollen grain is carried by the wind or a pollinator to the stigma of a flower of its own species, it germinates and grows into a long thin tube. The tube grows steadily toward an ovule inside the ovary of the flower, and if its growing tip survives long enough to reach that far-off goal, one of its nuclei fuses with an egg nucleus there. Those fused nuclei grow into an embryo, a miniature partly-formed plant that lies dormant inside the seed until conditions are right for germination.

productivity – A measure of the rate that living things capture energy. For example, desert plants capture about 2 kilocalories per square metre per day, while intensive year-round sugar cane agriculture captures 100. Productivity usually is limited by shortages of water or else nutrients, especially nitrogen and phosphorus. However, whatever is in shortest supply relative to the plant's requirements will be the most important "limiting factor", and must be removed before the plant's productivity will increase.

promiscuous – Said of flowers that attract a great variety of pollinators rather than only a specialized few. The pollinators themselves also are said to be promiscuous.

root-parasite – A plant that parasitizes other plants growing near it by plugging into the conducting tissues of their roots (see **haustorium**). If the parasite is green and makes its own food by photosynthesis in spite of being a root-parasite, it gets the slightly less disapproving label of *green* root-parasite. Many eyebrights (*Euphrasia*) here and abroad are green root-parasites, as are their relatives cow-wheat (*Melampyrum*), lousewort (*Pedicularis*), Indian paint-brush (*Castilleja*), and witchweed (*Striga*). The hosts of most root-parasites suffer little harm, but in recent years, witchweed has devastated millet crops in drought-stricken parts of Africa, even forcing farmers to abandon their land.

stigma – The pollen-catching surface of a carpel or ovary, often at the tip of a style. The pollen germinates on the stigma, and the pollen tube grows down through the style to the ovary, where fertilization occurs.

stomate (<u>or</u> **stoma**, plural **stomata**) – Stomates are tiny pores in plant surfaces. Most are on the underside of leaves, where they total about 1% of the surface. In some plants they are sunk into pits or channels or are covered with thick hair. Carbon dioxide (a raw material for the plant's food-making photosynthesis) can diffuse into the plant only when the stomates are open, but water vapour (vital as a plant stiffener as well as another raw material for photosynthesis) can escape from the plant at the same time. Thus the plant risks wilting whenever it makes food. However, that conflict is resolved automatically by the control system that opens and closes the stomates. The control system really is *two* controls – the first control opens or closes the stomates depending only on how much carbon dioxide is inside the leaf, and the second control *over-rides* the first if water gets scarce inside the leaf. That dual control system is able to provide food yet prevent thirst, because it responds to what is actually happening inside the leaf rather than to outside conditions like light, temperature, wind, and soil moisture.

style – The stigma-bearing tip of a carpel or ovary. Styles come in diverse forms (at least a dozen forms have been given their own names). Also, the style count can range from zero to many.

system – Anything that changes an input to an output. A system can be machinery like an electric drill and bit, but it can be biological too. The input goes by various names depending on the system, names like force or stimulus (the input of the drill is electrical energy). The output gets various names, too, like effect or reaction (the output of the drill is a hole). Some systems can handle different inputs. Your gut is like that – the enormous variety of foods that you eat is broken down to only a few basic molecules that you build into your own body or burn as fuel. Other systems can be altered to produce a new output from the same input. Most modern electric drills are like that – they can be changed into impact drills at the press of a button.

   Because the system and its input and output are all related, you might suppose that you could figure out any one of the three by knowing the other two, but that can be tough to do if you happen to know the input and output and hope to figure out the hidden details of the system itself. Systems analysts call that the "black box" problem. Molecular biologists over the last half century have enjoyed dazzling success figuring out in intimate detail the gooey guts of the many mysterious black boxes inside living things.

tomentum – A dense covering of matted woolly hair on a plant's surface.

weather (verb) – To break up rock into smaller and smaller bits. Rocks are *physically* broken up by root-growth and by freeze-thaw cycles. They are also *chemically* decomposed by solution, hydration, carbonation, and oxidation. For example, the bizarre karst topography of huge caverns, sink-holes (collapsed caverns), and disappearing rivers in our limestone mountain areas is a result of the slow but steady loss of calcite from the limestone, first chemically attacked by water and carbon dioxide and then literally washed away as soluble calcium bicarbonate.

An alpine forget-me-not, *Myosotis monroi*.

The insect-catching leaves of a sundew, *Drosera spathulata*.

A single leaf-hair of *Drosera spathulata* with its blob of sticky glue.

index

Acaena anserinifolia 77.
Adenochilus gracilis 7.
Anisotome pilifera 12.
astelias 116.
bellflower family 119.
blue-green algae 19, 20, 21, 22, 24.
borage family 2, 75, 76.
Boraginaceae 2, 75, 76.
Brachyglottis adamsii 100.
breeding barriers, 95, 96, 98, 104, 105, 107, 108, 110.
bristly mountain carrot 12.
Bulbinella hookeri 31.
buttercup family 45, 48, 124.
Caladenia lyallii 101.
calcium oxalate 83, 88.
camouflage 74, 76.
Campanulaceae 119
carbon dioxide 35, 52.
carrot family 12.
Celmisia dallii 126.
celmisias 116
Celmisia traversii 35, 39.
cephalodia 22, 24.
Cheesemania latisiliqua 6.
Chionochloa australis 49, 51, 52.
Chionochloa pallens 49, 52.
chromosomes 104.
Cobb Valley 14, 100, 107.
competition 8, 55, 57, 60, 61, 65, 70, 93.
Compositae 33, 35, 38, 53, 65, 68, 70, 74, 100, 119, 126.
Coprosma perpusilla 113, 114.
coprosmas 113, 114, 116.
Cruciferae 6.
crystals 7, 79, 83, 86, 88, 89, 90.
cultivars 96, 105.
cushions 8, 9, 61, 63, 65, 68, 70, 76.
daisy family 33, 35, 38, 53, 65, 68, 70, 74, 100, 119, 126
daphne family 63, 111.
divaricating shrubs 41.
Dolichoglottis scorzoneroides 53.
Donatiaceae 9.
donatia family 9.
Donatia novae-zelandiae 9.
Dracophyllum longifolium 86.
dracophyllums 81, 116.
Drapetes see *Kelleria.*
Droseraceae 16, 17, 18.
Drosera spathulata 16, 17, 18

edelweiss 33.
elaeocarp family 41.
Epacridaceae 39, 86, 121.
epacrid family 39, 81, 86, 121.
Epacris 86.
Epacris alpina 121.
Epacris pauciflora 86.
Epilobium glabellum 88, 89, 90.
Epilobium vernicosum 43, 44, 115.
Ericaceae 56, 105.
Euphorbiaceae 33.
Euphrasia cheesemanii 8.
Euphrasia laingii 27, 29.
Euphrasia monroi 26.
Euphrasia zelandica 72.
evening-primrose family 43, 44, 88, 89, 90, 115.
evolution 93, 116.
eyebrights 8, 26, 27, 28, 29, 30.
figwort family 8, 26, 55, 72, 74, 95, 96.
flaxes 119.
forget-me-nots 2, 75, 76.
Forstera tenella 60.
frost 11, 41, 68, 116.
fungi 19, 38.
Gaultheria crassa 56.
gene pool 96.
Gentianaceae 107.
Gentiana sp. 107.
Gentiana tenuifolia 107.
gentian family 107, 119.
geraniums 119.
Geum leiospermum 79, 110.
Gramineae 49, 51, 52, 108.
grass family 49, 51, 52, 108, 113.
green algae 21, 24.
Haastia pulvinaris 70.
Haastia sinclairii 74.
Haematomma babingtonii 57.
hairs 2, 16, 17, 18, 35, 38, 39, 51, 52, 70, 76, 77, 79.
harebell 119.
haustoria 27, 28, 29.
heath family 56, 105.
Hebe ochracea 95.
hebes 116, 119.
Hebe topiaria 55.
Hoary Head 31, 41.
humus 14.
hybridization 105, 115.
Hypericaceae 91.
Hypericum japonicum 91.
ice-blast 41, 60, 61, 65, 70.
isolating mechanisms 98.
Kelleria sp. 63.
Lake Cobb 4.
Lake Peel 14.
Lake Sylvester 4.
Leucogenes grandiceps 33.

Leucopogon suaveolens 39.
lichens 11, 19, 20, 21, 22, 24, 57.
Liliaceae 31.
lily family 31.
Lobeliaceae 117.
lobelia family 117.
madder family 41, 113, 114.
Melicytus alpinus 57.
mistletoe 30.
moas 41.
Montia calycina 93.
Mt Arthur 6, 12, 14, 21, 33, 38, 45, 49, 52, 116.
mustard family 6.
Myosotis angustata 75, 76.
Myosotis monroi 2.
natural selection 48, 93, 101.
nitrogen 7, 8, 11, 12, 14, 16, 19, 20, 21, 22, 24, 26, 30, 31.
Northwest Nelson State Forest Park 4.
Nostoc 19, 20, 22, 24.
nutrients 4, 7, 11, 12, 16, 31, 53, 86, 93.
Olearia lacunosa ssp. *lacunosa* 38.
Onagraceae 43, 44, 88, 89, 90, 115.
Orchidaceae 7, 101.
orchid family 7, 101.
Oreoporanthera alpina 33.
oxalic acid 81, 83.
Oxalidaceae 81, 83.
Oxalis magellanica 81, 83.
pakihi 86.
Parahebe cheesemanii 74.
parasitism 8, 26, 27, 28, 29, 30, 57.
peat 14.
Pernettya macrostigma 105.
phosphorus 7, 11, 14, 16, 31.
photosynthesis 35, 52.
Phyllachne colensoi 61, 65, 68.
Pimelea longifolia 111.
Pittosporaceae 41.
Pittosporum anomalum 41.
pittosporum family 41.
Podocarpaceae 98.
podocarp family 98.
Podocarpus nivalis 98.
pollination 60, 98, 104, 107, 108, 110, 111, 113, 115, 117, 119, 121, 124.
porcupine shrub 57.
Portulacaceae 93.
portulaca family 93.
Potentilla anserinoides 4.
Pratia macrodon 117.
predators 2, 5, 6, 7, 16, 17, 18, 41, 65, 70, 72, 74, 76, 77, 79, 81, 83, 86, 88, 89, 90, 91, 93.
promiscuity 110, 115.
Pseudocyphellaria cinnamomea 20.
Pseudocyphellaria crocata 19.
Pseudocyphellaria homoeophylla 21, 22.
Psoroma pholidotoides 24.
Ranunculaceae 45, 48, 124.
Ranunculus foliosus 124.

Ranunculus insignis 45, 48.
Raoulia eximia 65, 68.
Rosaceae 4, 77, 79, 110.
rose family 4, 77, 79, 110.
Rubiaceae 113, 114.
Rytidosperma setifolium 108.
sandalwood 30.
scree slope 11.
Scrophulariaceae 8, 26, 27, 28, 29, 30, 55, 72, 74, 95, 96.
self-fertility 115, 117.
snow-berries 56, 105.
snow grasses 26, 49, 51, 52.
sphagnum moss 14.
spurge family 33.
St. Johnswort family 91, 119.
stomates 29, 30, 33, 35, 38, 39, 43, 44, 45, 48, 51, 52.
structure 2, 4, 6, 12, 22, 26, 30, 39, 51, 77.
Stylidiaceae 60, 61.
stylidium family 60, 61.
sundews 16, 17, 18, 19.
sunlight 4, 6, 43, 44, 45, 52, 53, 55, 57, 60, 61, 63, 65, 91.
Thymelaeaceae 63, 111.
tomentum 22, 38, 39.
tree daisy 38.
Umbelliferae 12.
Violaceae 41, 57.
violet family 41, 57.
Wahlenbergia marginata 119.
water 7, 9, 11, 14, 16, 29, 30, 31, 33, 35, 38, 39, 41, 43, 44, 45, 51, 52, 53, 70, 86, 93, 116.
willowherbs 43, 44, 45, 81, 88, 89, 91, 96, 115, 116, 117.
wind 4, 7, 33, 35, 38, 39, 41, 51, 52, 60, 61, 86, 108, 111, 113, 116, 126.